Fine Homebuilding

TRICKS OF
THE TRADES:

JIGS, TOOLS
AND OTHER
LABOR-SAVING
DEVICES

Fine Homebuilding

TRICKS OF THE TRADES:

JIGS, TOOLS AND OTHER LABOR-SAVING DEVICES

The Taunton Press

Cover Illustration: Robert La Pointe

Taunton
BOOKS & VIDEOS
for fellow enthusiasts

First printing: March 1994
Printed in the United States of America

A FINE HOMEBUILDING Book
FINE HOMEBUILDING® is a trademark of The Taunton Press, Inc.,
registered in the U.S. Patent and Trademark Office.

The Taunton Press, 63 South Main Street, Box 5506,
Newtown, CT 06470-5506

Library of Congress Cataloging-in-Publication Data

Tricks of the trades. Jigs, tools, and other labor-saving devices.
       p.   cm.
   At head of title: Fine homebuilding
   "A Fine homebuilding book"—T.p. verso.
   Includes index.
   ISBN 1-56158-076-7
   1. Building — equipment and supplies — Miscellanea.   I. Fine
homebuilding.   II. Title: Jigs, tools, and other labor-saving devices.
   TH915.T75   1994   93-48878
   690'.028 — dc20   CIP

CONTENTS

INTRODUCTION

Have you ever faced a seemingly simple problem on a project and not been able to come up with a solution? It's maddening, isn't it? You know there's an answer just beyond your reach. If you could just get your brain cranked up and focused, you'd be sure to get it.

Well, sometimes you do get it—and what a great feeling that is! But sometimes you don't, and you may wind up wasting time and materials in a solution that you later realize made no sense.

For a generation now, *Fine Homebuilding* magazine readers have had a forum for dealing with such intractable problems. Their solution has been two extraordinarily popular columns that appear in each issue: Tips & Techniques and Q & A.

Tips & Techniques is a monologue; it provides answers only. It is the filtered collection of hard-won tricks developed by experienced professionals and skilled amateurs facing real building problems. Hundreds are submitted to the editors, but only the select few are chosen to be published. And readers are constantly telling us how useful these ideas have been to them.

Q & A is a dialogue. Builders with problems they can't solve write in to the editors, who find pros who have faced and solved the same problem. Sometimes their response elicits other responses, and a fascinating exchange of ideas may ensue.

The net impact of both columns is the same: people sharing problems and solutions in an open forum and helping each other become better craftsmen.

This book, one of a set, is a collection of the best ideas from these columns published in the last five years or so. It focuses on subjects near and dear to the hearts of many do-it-yourselfers and professionals—building jigs, making new use of old parts, finding additional uses for tools, improving tool techniques, and creating devices out of items you thought were junk or just used for something else. It proves something that many of us have suspected for a long time—that the real pleasure is in the process and the discovery as much as in the satisfaction of completing a job well done.

1

JIGS, VISES AND CLAMPS

PIPE-CLAMP HOLD-DOWN

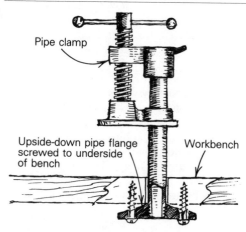

Pipe clamp

Upside-down pipe flange screwed to underside of bench

Workbench

Whenever I need a hold-down clamp on my bench, I reach for one of my pipe clamps. As shown in the drawing above, I insert a short length of pipe through a hole in the bench top. It is threaded into a pipe flange secured upside-down to the underside of the bench, giving me the versatility of a hold-down without having to buy one.

—*Joseph Kaye, Uniondale, N. Y.*

MULTIPURPOSE DOORPULL JIG

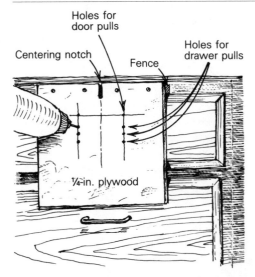

Holes for door pulls

Centering notch

Fence

Holes for drawer pulls

¼-in. plywood

This drawing shows a jig I use for quickly locating the screw holes for cabinet pulls. In the application illustrated here, the jig is being used on a drawer front. The notch at the top of the jig is aligned with a pencil mark on a piece of tape that indicates the center of the drawer. Drawers of different depths require their pulls to be placed at different distances from their top edges—hence the series of holes.

To use the jig for locating pulls on cabinet doors, I rotate the jig 90° and align its edge with the door's top edge or some molded detail in the door. The jig is laid out with equal distances from its sides to the pull holes, allowing it to be flipped to do right-hand or left-hand doors.

—Mark Hallock, Capitola, Calif.

HONING GLASS

In John Lively's excellent treatise on grinding and honing edge tools *(FHB #18, pp. 56-61)*, he argues for more skill and fewer gadgets. Lively mentions using wet/dry sandpaper for honing, but does not suggest using a hard surface as a flat for the sandpaper. For years I've honed edges with mixed results, until I switched to a glass mounted on a wood base.

My glass is ¼ in. thick by 6 in. sq. The glass shop charged $3.50 for it, which included polishing the edges. The ¾-in. pine base is 6½ in. by 12 in., with three birch dowels through the width to minimize warping. Four slender cleats at one end of the base frame the glass, without protruding above its surface. The remaining half of the base provides purchase for clamps.

I painted the base with three coats of oil-based enamel, and then set the glass in its frame with silicone adhesive. This makes the base waterproof and easy to clean under the tap.

To sharpen my edge tools, I cut pieces of 240-, 320- and 600-grit wet/dry sandpaper, and use them as seems appropriate to the edge I need. Used wet, the paper stays on the glass quite well, and I often use it on my kitchen counter next to the sink. For polishing an edge, I've found that auto rubbing compound (both red and white) also works well on the glass.

—Norin A. Elfton, Lakewood, Colo.

ADJUSTABLE CABINET JACK

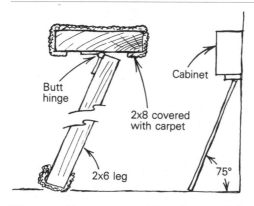

Butt hinge

Cabinet

2x8 covered with carpet

2x6 leg

75°

Whenever I hang upper cabinets by myself, I use a hinged cabinet jack like the one shown in the drawing above . It consists of a 54-in. long 2x6 leg with a 20-in. long 2x8 block attached (to the top end) by way of a 4-inch butt hinge. Both the block and the base of the leg are covered with carpet to prevent scratches to wall and floor and to make it easier to scoot the bottom of the leg toward the wall.

To use the jack, I position the block a little lower than the eventual location of the cabinet bottom. Then I load the cabinet atop the block and tap the leg toward the wall to raise the cabinet into position. The butt hinge allows the block to maintain complete contact with the cabinet as the leg's angle changes.

—Robert Francis, Napa, Calif.

CLAMP-FISHING

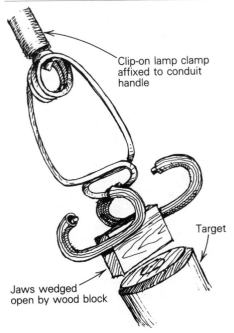

Clip-on lamp clamp
affixed to conduit
handle

Target

Jaws wedged
open by wood block

I dropped my hammer down a narrow space, well out of reach. I couldn't hook it because it landed head down. Looking around for a helpful tool, I noticed the clip-on light fixture that I was using. I quickly unscrewed the simple ball-and-socket joint to remove the spring clamp from the light, and then press-fit the ball into a length of electrical conduit that was handy. Then I opened the jaws of the clip and inserted a small block of wood to keep the jaws wedged open wide enough to fit over the hammer's handle, as shown.

I lowered this apparatus into the crevice and lightly placed the block against the handle. Then I gave a quick push. As the block fell away, the clamp closed firmly around the handle. I carefully extracted the hammer, and in a few minutes the light and the hammer were back in action.

—Brian Carter, Concord, N. H.

DECKING PERSUADER

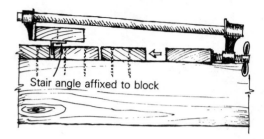

Stair angle affixed to block

Whenever I have to lay deck boards that have at least a ⅛-in. gap between them, I use a pipe clamp and a block to muscle the crowned pieces into place. The key to this system, as shown in the drawing above, is the steel stair-tread angle-bracket mounted on the block. With the projecting leg of the angle slipped into the space between a pair of secured deck boards, the block becomes a handy purchase for one end of a pipe clamp. Tighten the clamp to pull the crowned plank into place. Stair angles (such as the Simpson TA 10) are available from suppliers that carry metal framing hardware, such as joist hangers and post bases.

—Gregory Tolman, Mammoth Lakes, Calif.

SITE-BUILT BAR CLAMP

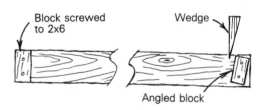

Block screwed to 2x6

Wedge

Angled block

When I found myself in need of a few extra bar clamps, I took a tip from the flooring profession, a line of work with practitioners who are constantly using wedges to force together pieces of wood. I screwed a wood block to both ends of a relatively straight 2x6 (drawing above). One block is perpendicular to the edge of the 2x6, the other at a slight angle. Then I cut a wedge to match the angle of the second block, and had an instant bar clamp.

—Chris Petersen, Storrs, Conn.

SHIM-SHINGLE JIG

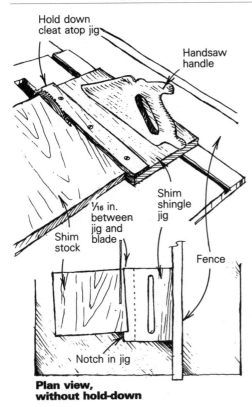

Hold down cleat atop jig

Handsaw handle

Shim stock

1/16 in. between jig and blade

Shim shingle jig

Fence

Notch in jig

Plan view, without hold-down

Here's a table-saw jig that you can use to make shim shingles from scraps of wood. The base of the jig is a rectangular piece of plywood, with an angled notch cut out on the side that passes by the sawblade. The jig's handle is also wood, patterned after that of a handsaw. To make sure the shingles stay in the jig's notch while cutting them, I screwed a hold-down cleat to the top of the jig, as shown in the top portion of the drawing above.

In use, I make a couple of passes with a piece of 1x shim stock and then flip the stock around to compensate for the tapered edge that's left over. With this jig I can quickly make shims that are more accurate than ready-made cedar shims.

—*John Kraft, Oakland, Calif.*

LEVELING ROD

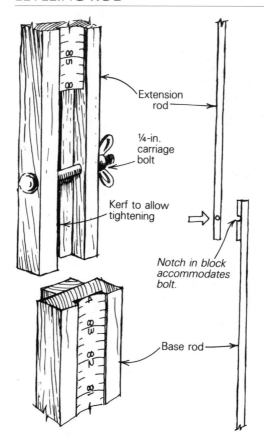

Extension rod

¼-in. carriage bolt

Kerf to allow tightening

Notch in block accommodates bolt.

Base rod

The drawing above illustrates an inexpensive design for an accurate, two-piece leveling rod made from a defunct 1-in. wide tape measure. To make the base rod, begin with a piece of straight, knot-free 1x stock, ripped to a full 2-in. width. I think 7 ft. is a good length for this section of the rod. Use a router or a dado head to cut a ⅞-in. wide by ¼-in. deep groove in the base rod. Use tin snips to cut a 7-ft. section of tape, and press-fit the tape into the groove. It should stay put.

To make an extension rod, cut another groove in a similar piece of wood, an inch or so shy of 7 ft. long. At one end, deepen the groove to ½ in. for the first 5 in. of the rod and cut a saw kerf 5 in. long in the center of the groove. Next, drill a hole for a ¼-in. carriage bolt, as shown in the drawing above . Now you can press the continuing portion of the tape into the groove, starting at the end of the kerf.

At the top of the base rod, glue a ⅞-in. wide block, about 5 in. long and ½ in. thick. This block has to be notched to accommodate the bolt in the extension rod. To use the extension, press the block at the top of the base rod into the groove in the extension rod. Adjust the extension up or down until its tape is even with the top of the base rod, and tighten the wing nut.

When the extension rod is not in use, I turn it upside-down and reattach it to the base rod, out of the way. This rod is accurate, and I think it's easier to read than commercial ones, which cost $60 or more.

—Jim Reitz, Towson, Md.

SANDING BLOCK

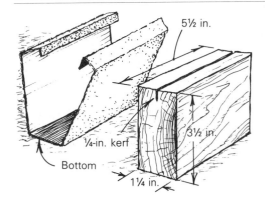

You can buy sanding blocks at the hardware store, but I've always made my own. All it requires is a small scrap of hardwood and a couple of passes with a saw. If the block is carefully cut and the sandpaper is crisply folded, the paper will stay snugly in place without gluing or tacking it.

Cut a block 1¼ in. by 3½ in. I use this width and height because the block feels comfortable in my hand. Cut the block 5½ in. long, the same length as half a standard sheet of sandpaper. A ¼-in. deep kerf along the center of one of the long edges completes the block.

As shown in the drawing above, there are six folds to be made in the sandpaper. Begin by inserting one edge of a half-sheet in the kerf. Fold the paper around the surface of the block until you reach the sixth fold. Gauge this crease by removing the first edge of the paper from the kerf, and inserting the second edge.

Once the sandpaper is all folded, hold the edges that will sit in the kerf together so that the sandpaper forms a sleeve, and feed it onto the block from one end.

—M. Felix Marti, Monroe, Ore.

WEDGING A MORTISE

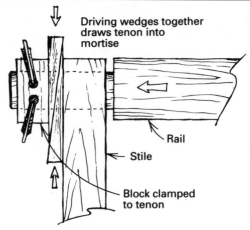

Driving wedges together draws tenon into mortise

Rail

Stile

Block clamped to tenon

While building a wide sliding barn door, I didn't have pipe clamps long enough to pull the rails and the stiles together. So I used the rails as clamps, as shown here. First I cut the rail tenons about 8 in. longer than they needed to be. Then I ran the tenons into their mortises as far as they would go without clamping pressure and clamped blocks to their ends. Opposing wedges driven between the blocks and the stiles brought the door frame together.

—*Ed Good, Nordland, Wash.*

STOVE-BOLT CLAMPS

Stove bolts and wing nuts

Hardwood jaws

Sometimes I don't have enough store-bought clamps to get me through a big glue job when I'm working with wood laminations. The little clamps shown in the drawing are an inexpensive solution. I rip strips of hardwood to suitable lengths and then drill holes near their ends. Coupled with stove bolts and wing nuts, the wood strips make quick, inexpensive clamps.

—*Donald E. Russell, Bolton Landing, N. Y.*

A WATER COUNTERBALANCE

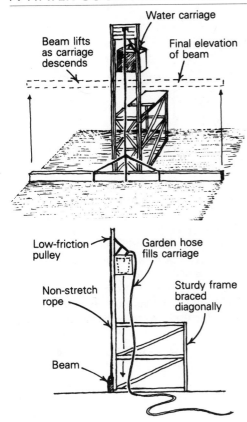

By constructing a crane with a water lift (as roughly diagrammed in the drawings above) one can lift a heavy load with relative safety. In addition to constructing a sturdy crane for the job, it's crucial to use a pulley that has a low coefficient of friction. We put a couple of plastic garbage cans in the carriage to hold the water.

As the water carriage is filled with water from a garden hose, it starts to lift the load on the opposite side of the pulley. Once its weight becomes greater than the weight of the load and it overcomes the friction created by the pulley, the carriage goes down and the load goes up.

I came upon this design when a foreman of mine wanted us to lift a beam weighing approximately 1,000 lb. 18 ft. above our heads, on ladders, with no safety lines. With the proper counterbalance, not only are the lifespans of backs, and perhaps careers, lengthened, the water does most of the work.

—*Max Wolf, Berkeley, Calif.*

CLAMPING HANDRAILS

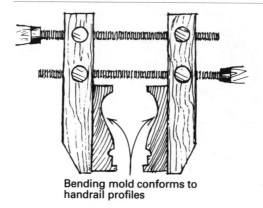

Bending mold conforms to handrail profiles

Whenever I have to join two sections of handrail together with rail bolts, I use a handscrew with pieces of bending mold screwed to the jaws to align the sections. As shown above, bending mold is milled to conform to the profile of the handrail, only in reverse. The type I've used is made by the L. J. Smith Company, Inc. (Rt. 1, Bowerston, Ohio 44695; 614-269-1133). They make bending mold in both yellow pine and polyester resin. The material is sold by the foot through their retail outlets.

—Mike Nathan, Hailey, Idaho

ADJUSTABLE CUTOFF FIXTURE

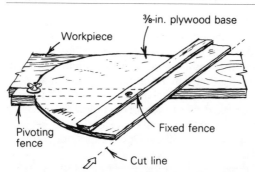

⅜-in. plywood base

Workpiece

Pivoting fence

Fixed fence

Cut line

I had to cut some acute angles on 11-in. wide, 16-ft. long oak planks. Rather than haul them back to the shop to use the radial-arm saw, I devised the cutting fixture shown in the drawing above. It's basically a big version of the metal saw protractors I'd seen in mail-order catalogs, but the materials were practically free, and I didn't have to wait for the mail carrier to bring it.

The ⅜-in. plywood base is 32 in. long. I made the distance from its right edge to the fence a little wider than the distance from the edge of my saw's base to the blade. I used a fine-tooth blade to trim off the extra plywood with my circular saw, using the fence to guide the cut. That made the right edge of the base parallel with the fence. Cuts are aligned on this edge.

The circular portion of the plywood base has a 13½-in. radius. A 1x2 fence that pivots on a machine screw is mounted under the fixture. I covered the edge of the pivoting fence that bears against the workpiece with a strip of sandpaper to keep the jig from slipping off its layout line. Once I've got the angle between the pivoting fence and saw-guide where I want it, I secure the fence to the base with a wingnut over a big washer.

—Brad Schwartz, Santa Ana, Calif.

SITE-BUILT WIRE SPINNER

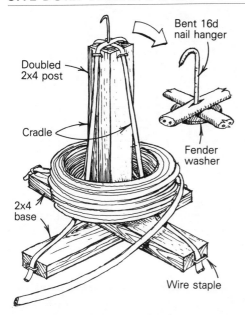

Workmanlike wiring is easier to achieve with a wire spinner. The site-built version shown in the drawing above was whipped up by electrician Phil Clements in about 15 minutes, using a handful of wire staples, a few 16d nails, a fender washer, assorted 2x4 offcuts and some short lengths of nonmetallic sheathed electrical cable (Romex). Phil first nailed together a pair of 2x4s to make a post about 24 in. tall, and then attached the 24-in. long base pieces. He next stapled the short lengths of wire cable to create a loose cradle that holds a coil of wire as it comes from the box. Hung from a nail in a ceiling joist or door header, Phil's wire spinner rotates on the washer as he pulls and uncoils flat lengths of wire without twists or kinks.

—*M. Scott Watkins, Arlington, Va.*

PIPE-CUTTING JIG

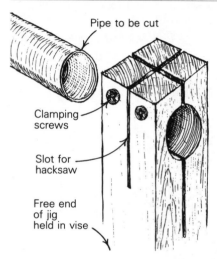

Pipe to be cut

Clamping screws

Slot for hacksaw

Free end of jig held in vise

When I had to repair some plumbing pipes under my kitchen sink, I found that one of the thin-walled waste lines had been trimmed with a tubing cutter too close to its end, resulting in a deformed, crimped cut that caused a leak. To make a new cut near the pipe's end without deforming it, I improvised the jig shown in the drawing above. First I drilled a hole the same diameter as the pipe in a piece of wood. Then I used a bandsaw to make a pair of intersecting kerfs. The kerf parallel to the pipe allowed the jig to be tightened on the pipe by way of a couple of screws. The other kerf provided a slot for my hacksaw blade. I held the jig fast with a vise as I made my cut, which required only a slight dressing with a file to deburr its crisp new end.

—*V. A. Maletic, Antioch, Calif.*

OUTLET JIG

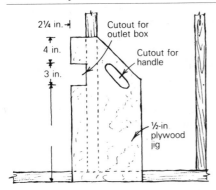

I use a jig to locate my electric outlet boxes in a stud-framed house. As shown in the drawing above, my jig is made out of ½-in. plywood. To use it, I place the jig on the floor against the studs, fit a plastic outlet box in the cutout and nail the box to the stud. The jig holds the box, shims it out so that it will be flush with the drywall and sets all of the boxes at exactly the same height. It cuts my rough-in time plenty.

—Norman Cook, Cockeysville, Md.

SCARF-JOINT JIG

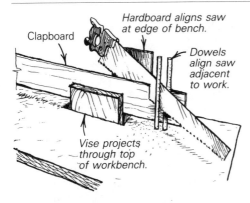

On my old house, some of the clapboard is still secured to the framing with square nails. Where clapboards abut one another, they are scarf-jointed. This house was built before the use of building paper, and clearly the overlapping scarf joints allowed less water infiltration. When it came time to repair some of the clapboards, I wanted to duplicate the scarf joints, which are cut at 20°. I couldn't find any information on how to make these cuts, so I devised the jig shown in the drawing on p. 21.

One of my shop benches has a vise that projects through the top. I use it to secure the siding, which is shimmed with an inverted scrap of clapboard to keep it vertical. I use a crosscut saw with a pointed end to make my cuts because I find that it's pretty easy to steer. A piece of hardboard screwed to the front of the bench aligns the saw on the near side of the work. On the far side, a pair of dowels let into the bench top keeps the saw in line.

—Guy Campbell, Norway, Maine

SITE-BUILT SHEET-METAL BRAKE

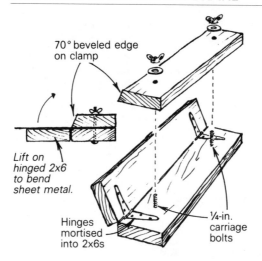

70° beveled edge on clamp

Lift on hinged 2x6 to bend sheet metal.

Hinges mortised into 2x6s

¼-in. carriage bolts

I needed a sheet-metal brake to bend some flashings, so I used some materials that were on hand to make it quick, simple and cheap. I began by cutting an 8-ft. 2x6 into three pieces of equal length. As shown in the drawing above , I let a pair of strap hinges into mortises that I cut with my router toward the ends of two of the pieces of 2x6. Using my table saw, I took a 20° rip off the third piece, leaving me with a 70° beveled edge. This beveled piece (let's call it the clamp) is affixed to the body of the brake with ¼-in. carriage bolts that are 3½ in. long.

To bend a piece of sheet metal with the brake, I slip the sheet between the clamp and the neighboring 2x6. Then I tighten the wingnuts and lift up on the hinged 2x. Result: a nice, crisp bend.

—William Boyce, Mukilteo, Wash.

DRILLING SHEET METAL SAFELY

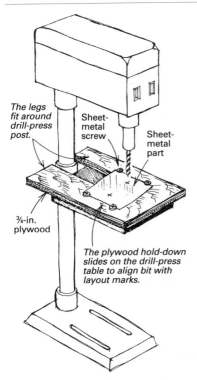

The legs fit around drill-press post.

Sheet-metal screw

Sheet-metal part

¾-in. plywood

The plywood hold-down slides on the drill-press table to align bit with layout marks.

Boring holes in a piece of sheet metal can present a drill-press operator with a potentially hazardous situation. Sometimes the piece won't fit in the drill-press vise, and clamps or Vise Grips often slip and can be cumbersome to use. Even worse, inexperienced operators will try to hold the workpiece with their bare hands. If the bit binds, the piece can spin around and gash a hand or arm with its sharp edges.

The drawing above shows a hold-down fixture that I use to anchor small pieces of sheet metal during drilling. As shown, it's a 20-in. square piece of ¾-in. plywood with a U-shaped notch cut out of it to create a couple of legs that fit loosely around the drill-press post. I position the part I want to drill on the plywood and secure it along the edges with four sheet-metal screws. In use, I slide the hold-down around the drill-press table as needed to center the bit on the center-punched points that need to be drilled. The oversize notch in the plywood makes it easy to move around the drill-press post. At the same time, the legs keep the hold-down from spinning out of control should the bit bind and the operator lose his grip on the hold-down.

—*Mark Francis, San Diego, Calif.*

2

HAND TOOLS

MAKESHIFT SOCKET EXTENSION

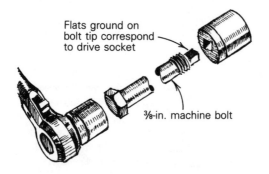

Flats ground on bolt tip correspond to drive socket

³⁄₈-in. machine bolt

An old guy had paid me good money to build a deck for him, but then he insisted on helping to make sure I did it right. At one point, while we were bolting the stair stringers to the posts, I discovered that I had placed a ½-in. bolt where it was difficult to reach with an ordinary wrench and a socket. It was down in a crevice between a couple of 2x6s, and neither of us had a long socket extension. We sat and pondered a bit, and then the old guy disappeared. In a few minutes he returned with one of our long ³⁄₈-in. bolts. He had ground the threaded end square so that it fit exactly in a ¼-in. drive socket (drawing above). As I tightened a nut onto the bolt, I looked back at the old guy. He smiled and gave me a wink.

—*J. Azevedo, Santa Clara, Calif.*

TIRE NOOSE

After several trips to the local gas station to get the tire fixed on my wheelbarrow, I learned a valuable lesson. The hard part of fixing the tire wasn't patching the hole—it was getting the tubeless tire to inflate once the bead was broken. The drawing above shows how to get the rim and the tire back together again without having to pay the service-station attendant to do it.

Tie a loop at the end of a piece of rope and wrap the rope around the tire. Put the running end of the rope through the loop, and pull snug on it. Once the rope is tight enough, the tire bead will be touching the rim, and you can use a compressor to inflate the tire. A little grease around the bead will also help you to get a good seal.

—Bruce Schwarz, Manchester, Md.

A BETTER TIRE NOOSE

Bruce Schwarz's tip on inflating a tubeless tire with the help of a rope noose is good advice, but a cloth strap clamp works even better. The wide, thin strap holds better on the treads, and the locking cam allows you to apply a lot of pressure without worrying about it releasing at the wrong time.

—J. Kaye, Uniondale, N. Y.

STEADY THAT BOB

If you have to use a plumb bob on a windy day, dangle it in a pail of water to dampen its sway.

—Joe Graczyk, Cazadero, Calif.

PLUMB-BOB TIP

When the tip to my plumb bob was missing, I stopped by the local lumberyard to pick up a new one. No such luck. I'd either have to wait two weeks for a special order, or buy a whole new bob. Faced with these alternatives, I went through the machine-screw bins until I found a screw that matched the threads of my missing tip. Then I cut off its head and ground a suitable point on it. The real bonus here is that I discovered I could use the long machine screw tip as a plumb-bob extension. By backing the tip out a bit, I can set its elevation at precisely the point I want.

—*Bruce Bieschke, Eureka Springs, Ark.*

SLEDGE BUMPER

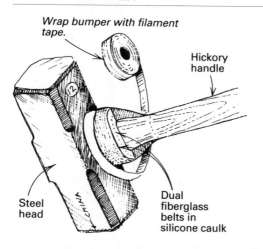

Wrap bumper with filament tape.

Hickory handle

Steel head

Dual fiberglass belts in silicone caulk

The hickory handle on my 12-lb. sledge is tough enough to last a lifetime, but not if the person using it overshoots the mark and bashes the target with the handle. Mistakes do happen, so I give my sledge an overstrike protector.

I begin by cleaning the handle and abrading it and the adjacent metal of the head with some heavy sandpaper. Then I apply overlapping beads of acetic acid-cure silicone caulk to the prepared section. As shown here, I lay it on thickest where the damage is most likely to occur. Then I cut a swath of fiberglass cloth as wide as the band of silicone and long enough to overlap itself by an inch and gently wrap it over the caulk. Smoothing the cloth with a finger or a spatula gives excellent shape-holding results and wets out the fabric. I repeat the process, then I apply an additional thin layer of caulk over the second layer of fabric and cover the whole assembly carefully with plastic wrap to make a smooth finish. The plastic wrap can be peeled off in a day, but the mass needs about a week to cure fully. I then wrap my bumper with a sacrificial layer of filament tape. The 3-year-old bumper on my 12-lb. sledge is like the day I made it, although the tape has been badly cut and torn.

—William H. Brennen, Boulder, Colo.

LOW-BUDGET WATER LEVEL

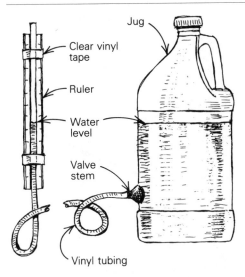

Jug

Clear vinyl tape

Ruler

Water level

Valve stem

Vinyl tubing

An inexpensive water level can be made out of a clear plastic 1-gal. jug, a tubeless-tire valve stem (with guts and cap removed) and an appropriate length of clear vinyl tubing, as shown in the drawing above . First, drill a hole that is sized to accept the valve stem approximately 2 in. up from the bottom of the jug. Insert the valve stem in the hole, fit one end of the tubing over the stem and tape a ruler to the free end of the tubing to be used as a reference stick. Finally, fill the jug with water and add a few drops of food coloring to make the level easier to read.

Before using the level, be sure to bleed any air bubbles out of the tubing. And during use, keep the free end of the tubing above the level of the water in the jug to keep water from draining out the free end of the tube. Water levels are very accurate, and this one can be just the ticket for someone who doesn't need one often enough to justify buying a commercially manufactured one.

—Jeff Jorgensen, Tonopah, Nev.

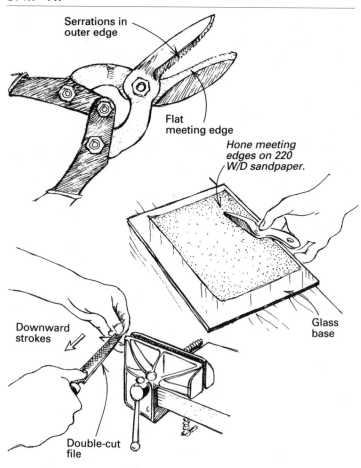

Serrations in outer edge

Flat meeting edge

Hone meeting edges on 220 W/D sandpaper.

Glass base

Downward strokes

Double-cut file

Tin snips get dull, just like every other edge tool in the builder's toolbox. But the right way to sharpen them isn't widely known. The drawings above show how I sharpen my aviation-type metal snips.

First I take the jaws apart by removing the screws that hold them together. Next I hone the meeting edges of the jaws on a piece of 220 wet/dry sandpaper placed atop a glass sheet.

The last step is to serrate the outer edges of the two jaws with a double-cut file. The trick here is to remember that the file will leave serrations when used in one direction but not the other. Use moderately hard, short, downward strokes to cut the teeth. In the drawing shown here, the file moves from right to left, or heel to toe. Reassemble the snips, and they're good as new—maybe even better.

—*Tom Law, Westminster, Md.*

HOLDING A LEVEL PLUMB

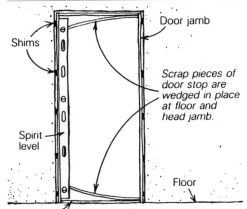

Door jamb

Shims

Scrap pieces of door stop are wedged in place at floor and head jamb.

Spirit level

Floor

Temporary spacer cut to exact opening size

I use a 6½-ft. spirit level to plumb and straighten interior door jambs. In order to leave both hands free to handle the shims and to drive nails, I wedge the level against one of the jambs to hold it in place. This lets me monitor the jamb continually as I make fine adjustments with the shims, rather than having to check and recheck the alignment with a hand-held level.

Once I've assembled all the jambs for the house, I cut a spacer the exact width of the door opening out of 1x6 or jamb offcuts. As shown in the drawing, the spacer rests on the floor between the jambs, maintaining the correct dimension at the bottom of the door opening.

Spacer in place, I temporarily shim the jamb in its rough opening. Then I wedge the level against one of the jambs with scrap pieces of door stop. The spaces above and below the level should be equal. Now I can plumb and straighten the jamb in the usual manner, tapping shims in and out until the jamb is straight up and down and true as an arrow. Then I nail the jamb to the jack studs with a pair of 10d finish nails through each set of shims. After double checking for alignment and making any necessary adjustments, I repeat the procedure on the opposite jamb.

—D. B. Lovingood, Portsmouth, Va.

TUNING AN ADJUSTABLE LEVEL

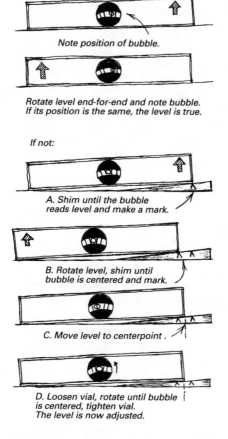

Note position of bubble.

Rotate level end-for-end and note bubble.
If its position is the same, the level is true.

If not:

A. Shim until the bubble
reads level and make a mark.

B. Rotate level, shim until
bubble is centered and mark.

C. Move level to centerpoint .

D. Loosen vial, rotate until bubble
is centered, tighten vial.
The level is now adjusted.

For a dose of unadulterated frustration, there's nothing like finding out that the level you used to frame up that last window opening was totally out of level. Nine times out of ten, you won't notice until you step back and see the tilt of the sill and header in relation to the other wall framing.

To prevent this scenario from happening to you, get into the habit of checking a level for trueness before using it. This is especially necessary if the level belongs to someone else, or if you suspect that it has been dropped.

The true check is a simple procedure. As shown above, the level is placed against a flat surface (it doesn't have to be level), and the position of the bubble is noted. The level is then turned end-for-end. If the level is true, the bubble will come exactly to the same place. This test also applies for checking the vial that reads plumb.

If the bubble isn't cooperating, the level is off. Adjust it by following steps A-D in the drawing. All you need is a screwdriver that fits the vial's set screw, a flat surface and a scrap of shim stock.

—Jim Tolpin, Port Townsend, Wash.

LITTLE PLUGS

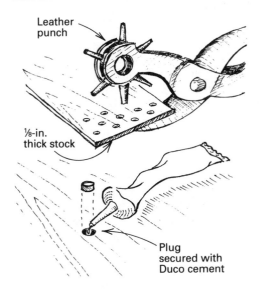

Leather punch

⅛-in. thick stock

Plug secured with Duco cement

I am a stairbuilder, and I use thousands of trim screws every year. I countersink the screws into ³⁄₁₆-in. dia. holes, then fill them with plugs made of the appropriate wood.

However, not one of the wood-shop supply outfits that I deal with has a source for ³⁄₁₆-in. plugs or plug cutters. While researching the subject, it occurred to me that I could use a leather punch to make small-diameter plugs (drawing above). This tool, which looks like a cross between a spur and a pair of pliers, has a rotating wheel full of dies designed to punch holes of varying diameters. I ripped a piece of wood to ⅛ in. thick and the punch spit out a perfect plug. When I install the plugs, I first put a dab of Duco brand cement (available at hobby shops) in the hole. Then I wet my finger, pick up a plug with it, insert the plug and drive it home. I sand it immediately with my palm sander, and I'm ready to move on to the next hole.

—Bob Johnston, Albuquerque, N. Mex.

SQUARE-TIPPED DRIVER

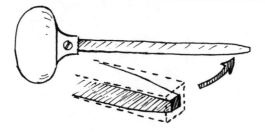

Working on a faucet seat recently, I came up against a fitting that required a square-tipped driver. I didn't have one that fit, but I did have an old doorknob with a square shaft. I used a grinder to shape one end of the shaft to the right profile, as shown in the drawing above, and I was back in business.

—Howard Furst, Bellingham, Wash.

NON-SKID TOOLS

It's exasperating and dangerous to have hand tools such as hammers, staplers or pry bars slide off a roof. To increase the likelihood that they'll stay put when you set them down, wrap their handles with layers of rubber bands that are thick enough to make positive contact with the roof deck.

—Betsy P. Race, Euclid, Ohio

3

POWER TOOLS

ROOF-GUN TUNE-UP

I use a pneumatic staple gun to install composition shingles, and every now and then the tool will clog up because roofing tar gets into the tip and the safety mechanism. Now I carry a can of Gum Out carburetor cleaner on the roof with me. Whenever the gun starts to get sticky, I give it a shot of the carburetor cleaner. It dissolves the tar without my having to disassemble the tool. Gum Out is available at auto-supply stores.

—Jeffery A. Oswalt, Kimmell, Ind.

NON-SKID NAIL GUN

As a roofing contractor, I have found that nailguns have a tendency to slide off roofs—even if the pitch is only 4-in-12. To keep mine from disappearing over the edge, I apply thick pads of silicone caulk to the parts of each gun that contact the roof when the gun is at rest. I'm not stingy with the stuff. The silicone pads should be at least 1 in. square and ½ in. thick. Not only do the pads provide traction for the tools on the roof, they also protect the metal casings during the course of daily use. It's not pretty, but it works.

—G. Walther, Tylersport, Pa.

RECYCLING WINDOW SHADES

Every time I buy an old house or work on one, it seems that a
big pile of old window shades gets canned. Don't throw those
shades away!

I just built a great outfeed table for my standing power tools out
of a bunch of 30-in. long by 1-in. dia. shade rollers. One end
already has a centered pin—a 16d nail works on the other. I put
several other shades to use as disappearing dust covers over my
drawing table and my hobby workbench, and the vinyl shades
have also found new uses as dropcloths.

—Frederick E. Bishop, Farrell, Pa.

SANDER SAFETY WARNING

I often use a belt sander to grind chisels and other pieces of metal.
After grinding the bevel on one of my chisels recently, I resumed
my work and noticed the smell of wood smoke in the air. Searching
for its source, I spotted small burn holes appearing in the sander's
cloth dust-collection bag. I immediately emptied the bag to find a
mass of smoldering wood dust ready to burst into flames.

This is just the kind of thing that can lead to a tragedy. If you use
your belt sander to grind metal, always remove the dust collection
bag before doing the work.

—Robin Ferguson, Snowmass, Colo.

CLEAN SANDING BELTS

Sanding the varnish off an old piece of furniture clogs a coarse-grit
sanding belt in a hurry. In this situation I use two belts. While I work
with one, the other sits in a small can of lacquer thinner. This
softens the gummy varnish enough that I can knock it off the belt
with a wire brush before reusing it.

—Steve Johnson, Washougal, Wash.

SITE-BUILT ROLLER TABLE

PVC pipe frame

PVC pipe roller

File flat.

On a recent job I needed some infeed and outfeed rollers to support the stock I was milling with my portable planer. I had a couple of Workmate® benches on hand, and some miscellaneous plastic drain pipes and fittings. By combining them, I came up with a pair of roller tables like the one shown in the drawing above. Most of the assembly is made of 1½-in. pipe. I used PVC, but ABS pipe would work just as well. The roller portion is a piece of 2-in. pipe. It fits over the 1½-in. pipe crossbar, and rolls freely as the stock passes over it.

The one modification I had to make to get this to work was to file a flat spot on the tops of the elbow fittings to keep them from protruding above the roller. Held fast by the bench's clamps, a pair of these rollers made a stable work surface that was easy to adjust up and down to compensate for irregular floors.

—Thomas Bujak, Palmyra, N. J.

SHOP-VACUUM NOZZLE

Duct tape seal
between bottle and hose

I do a lot of interior finish work. Sometimes the owners have
already taken up residence, and the dust and chips created by my
work can be a real nuisance. To gather up the mess that I make,
I've been using a plastic bottle as a nozzle. As shown here, I cut the
bottom off the bottle and angle the sides upward to resemble a
scoop. Affixed to the hose with some duct tape, the scoop lets me
easily pick up big and small debris at the same time.

—Michael Sweem, San Francisco, Calif.

VACUUM EXTENSION HOSE

On a recent remodel project I had to remove the wall and ceiling
sheathing over a stairwell. Before I started to install the new
finishes, I wanted to clean up the dust and cobwebs left on the
rafters. Because the area was above the stairs, it was tough to
reach the framing members—all my ladder setups were quite
precarious. My solution was to get a length of sump-pump
discharge hose. I bought a 24-ft. roll of it at the local lumberyard
for less than $5, and I taped it to the end of my shop-vacuum hose.
I affixed the end of the hose to a long furring strip, allowing me to
poke the hose into the dirty ceiling nooks and crannies from the
safety of the stair landing.

—Jeff Evers, Deforest, Wis.

TALKIN' TRASH

I line my shop vacuum with a plastic trash bag to make it easier to empty without generating the clouds of dust ordinarily associated with dumping a vacuum's contents. To keep the bag from obstructing the vacuum's filter, I place a disposable cardboard basket inside the bag. The cardboard basket should be cut to a height of no less than three-quarters of the canister height and should be of sufficient length to encircle at least one half of the canister's circumference. This technique will also boost the suction in a vacuum with a leaky canister.

—Mark Genovese, Taunton, Mass.

STORING SHOP-VACUUM TOOLS

The nozzles and hose for my wet/dry barrel-style shop vacuum were inconvenient to carry to jobs, and they tended to become separated from the main unit. I tried storing them in the barrel but this was a messy, slow procedure. My next refinement put an old bicycle inner tube to a new use. I wrapped it twice around the barrel, with the wraps 6 in. apart. This elastic waistband holds the accessories conveniently strapped to the machine.

—William H. Brennen, Boulder, Colo.

POWER-PLANE DUST BAG

At times I have to use my power plane in finished rooms, and without a dust bag the chips generated by the tool make a real mess. My plane came without a dust bag, so I made my own by cutting the legs off a pair of panty hose. I tucked one inside the other, and slipped the open end over the chip discharge port. Held in place by a sturdy rubber band, the bag works well.

—Theodore F. Haendel, Great Neck, N. Y.

AIR-HOSE REPAIR

No matter how well stocked a contractor may be, he is not likely to have more than one air hose per nail gun or jack hammer. A leak in that hose can therefore be as troublesome as it can be frequent. I offer a tried-and-true job-site repair of small punctures in high-pressure air hoses.

First, depressurize the hose. Wrap it firmly with very close wraps of form-tie wire, or better yet, 12-ga. or 14-ga. solid copper wire. The wraps should touch each other and extend 1-in. beyond the leak on both sides. It is important that the wire not be small enough in diameter or tight enough to cut the rubber hose when the hose later expands. Bend the cut ends of the wire over the wraps where they won't threaten to cut the hose and wrap the entire repair with duct tape to keep it from unraveling. When the hose is repressurized, air will force the walls of the rubber hose outward against the wire wrap and compress the puncture closed.

—Jan Lustig, Berkeley, Calif.

HORIZONTAL HOLES

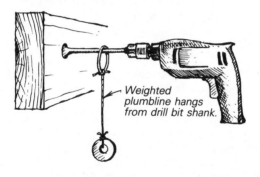

Weighted plumbline hangs from drill bit shank.

If you're having trouble drilling a horizontal hole with an electric drill, try hanging a plumb line from the bit's shank as you bore the hole. I use a brass key ring or a metal shower-curtain ring and let it ride on the shank. From it, I hang a string tied to a couple of washers (drawing above). When a spinning bit is held level, the weight will hang in place. Make sure the ring doesn't get hung up in the spinning chuck.

—Dave Herbert, Laurence Harbor, N. J.

OVERHEAD DRILLING

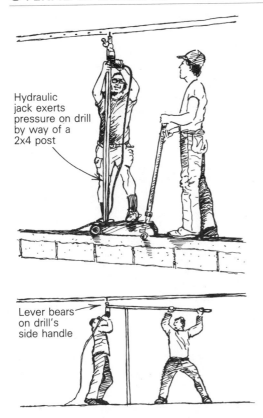

Hydraulic
jack exerts
pressure on drill
by way of a
2x4 post

Lever bears
on drill's
side handle

Even with a proper drill and a sharp bit, repetitive drilling overhead in concrete or steel can wear out your arm muscles. The two methods shown in the drawing above show how a helper and I solved the problem.

In the top half of the drawing, a hydraulic floor jack with a 2x4 post atop its lift plate supports the bottom of a drill. One person holds the drill and operates its switch, while the helper uses the jack-handle's lever to maintain constant pressure. With a little practice, the helper will learn how much pressure to apply just by the sound of the drill.

A simpler technique is illustrated in the lower portion of the drawing. Here a horizontal lever made of a length of pipe bears on a 2x4 post. A notch in the top of the post helps to center the lever. The lever pushes up on the side handle of the drill.

—Joseph Fetchko, Ocean City, Md.

TIGHT-SPOT ROUTERING

While applying plastic laminate to a bathroom vanity bordered by walls on three sides, I ran into trouble when it came time to make the cutout for the sink. Problem was, the base of my router was too wide to clear the back wall. So I removed my laminate trimming bit from the router, placed it in my ½-in. drill, and by holding the drill firmly and moving it slowly I easily completed the cutout.

—Scott Francoeur, Greenville, R. I.

ADJUSTABLE ROUTER BASE

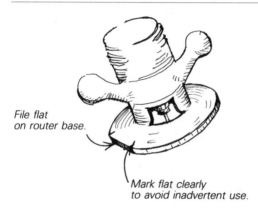

File flat on router base.

Mark flat clearly to avoid inadvertent use.

Sometimes when I'm using my router I find it necessary to make a cut a whisker wider than the jig or fence will allow. For instance, a dado might have to be a little wide to accommodate a piece of stock that is slightly oversize. Rather than go through the hassle of moving the jig or fence, which rarely yields satisfactory results, I rotate my router so that a flat portion of its circular base bears against the guide.

I used a file to create a flat spot on the edge of the router base, reducing its radius by about ⅟₃₂ in. at this point. By making a second pass on a cut with the flat spot against the guide, I remove another ⅟₃₂ in. of material from the workpiece. As shown in the drawing above, I clearly marked the flat spot on the base to avoid its inadvertent use.

—Wm. D. Holmes, Moscow, Maine

SITE-CUT PLUGS

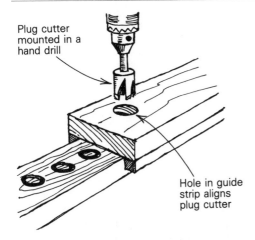

Plug cutter mounted in a hand drill

Hole in guide strip aligns plug cutter

If you want to get the best match with cross-grained plugs over counterbored screw heads, cut the plugs from the same wood as the work piece. This usually means they are best cut on site, but it's difficult to use a plug cutter in an electric drill held freehand, and most sites don't have the luxury of a resident drill press.

You can, however, use a plug cutter in a hand-held drill in conjunction with the guide strip shown in the drawing above . The hole in the guide corresponds to the outside dimension of the plug cutter. When clamped to the workpiece, the guide will keep the plug cutter from wandering or entering the work at an angle that is out of square with the work.

—Percy W. Blandford, Stratford-upon-Avon, England

CHUCK-KEY RING

It's annoying to misplace the chuck key for any drill, and dangerous if one is left in a drill press. I solved the problem in my shop by attaching my chuck key to a retractable key ring—the kind that is typically clipped onto a belt. I affixed the key ring to the side of my drill press. Now it's always at a convenient distance, and the key ring's spring-loaded cable always reminds me to remove the key when I'm finished using it. This type of key ring might also work well for hand drills—especially the cordless kind that is notorious for turning up keyless.

—Mark V. Genovese, Taunton, Mass.

CLAMPING SCARF JOINTS

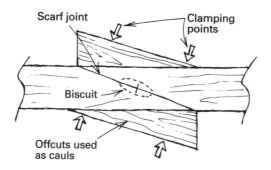

I enjoyed the article on table-saw jigs *(FHB* #53, pp. 58-61). The "board stretcher" scarf-joint idea has proved especially useful, but I've eliminated the assembly jig shown in the article to exploit another tool: the biscuit joiner. Using a biscuit ensures correct alignment, a strong joint and speeds up the joining process by eliminating the "one joint at a time in the jig" delay (drawing above). And as I explained to my wife, it gives me a good reason to have a $160 biscuit joiner and a rack full of clamps (the technique works better than the argument).

The sequence goes: cut the scarf joints and save the wedge-shaped offcuts; align the pieces and mark a centerline for registering the biscuit joiner; cut the slots, apply the glue and use the wedges as clamping cauls to distribute clamping pressure and to provide parallel bearing surfaces for the clamps.

—James A. Berg, Olympia, Wash.

CORD CONTROL

I tie an old boot lace or short length of heavy twine near the end of my extension cords. At the end of the day, I roll up a cord, wrap the two loose ends of the lace around the coil and tie them together. Then I've got a coiled cord that won't get snarled in the back of the truck.

—R. Francis, Napa, Calif.

RIGHT-ANGLE PLUGS FOR TOOLS

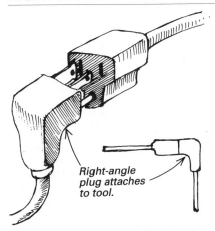

Right-angle plug attaches to tool.

It's always been frustrating for me when a power tool pulls loose from my extension cord. I've used twist-lock connectors, but they're bulky and hang up on things even more than a standard socket does. I've solved this problem by putting right-angle plugs on the cords to my tools, as shown in the drawing above. Because they're not on the same axis as the socket, right-angle plugs are a lot harder to pull out. These plugs are available at hardware stores with or without a ground connector. In my area the plugs cost between $1 and $7.

—Bob Francis, Napa, Calif.

4

SAWS

SAWBLADE JACKET

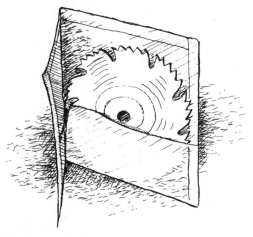

I get a lot of literature about building products. Much of it comes in heavy paper folders with pockets on the inside. After I've filed the literature in the proper drawer, I use the folders to protect my circular-saw blades. As shown in the drawing above, blades tuck neatly into one of the interior pouches, protecting their teeth. The pouches are usually 9 in. wide—big enough for almost all the blades I typically use. With the blades tucked in the folders, I can store them like books on a shelf.

—*Mark Feirer, Woodbury, Conn.*

HANDSAW SHOE

Rather than let my freshly sharpened handsaw fall victim to encounters with teeth-dulling objects in the back of the truck, I made a protective shoe for it out of a piece of 1x2. I cut a ½-in. deep kerf down the center of the 1x2 to house the saw's cutting edge. Then I drilled an ⅛-in. dia. hole about midway along the shoe for a wire hoop, as shown in the drawing above. The hoop wraps around the sawblade, holding the shoe in place. To snug the hoop to the blade, I twist the wire. Then I recess the cut ends of the wire into the wood with a nail set.

—Bob Jewell, Kalaheo, Kauai, Hawaii

AVOIDING SAW MARKS

When trimming doors or countertops, carpenters often apply some sort of tape to the base of their saws to keep from marring the work. Masking tape is a bad choice for this because it isn't very durable and it's tough to remove. Duct tape is also a poor choice because it makes the saw hard to push. Mylar tape, on the other hand, is slippery enough to glide over the work, and it's also easy to peel off the saw's shoe.

—Dennis Lamonica, Panama, N. Y.

GREASING BLADES

To make a clean cut with my circular saw, I get out the paraffin wax and rub it on the sawblade teeth (don't forget to unplug the saw). Whether it be an exposed beam or an expensive door with a delicate veneer, the wax helps the blade glide through the material. I find that a little wax on the teeth of a boring bit also helps when I'm faced with boring holes for locksets.

—Richard Rose, Cedar Grove, Wis.

SAWZALL BLADE LOCK

If you've got a Milwaukee Sawzall, put a small lock washer between the Allen-head screw and the clamp that anchors the blade. This minor improvement will keep the screw from loosening during use, saving you the trouble of mid-cut stops to tighten it and reducing wear on the Allen-head socket.

—Mark Handrahan, Hingham, Mass.

SAW HANGUP

1x2 affixed to sawhorse leg

The much-abused worm-drive saw often ends up lying in the mud next to the sawhorses between cuts. Not only is this hard on the saw, but the endless bending and stooping to put the saw down and pick it up again is also hard on the back.

One day I watched my brother staple a short length of 1x2 to the leg of his sawhorse to act as a saw hanger, and I've been doing the same ever since. This surprisingly handy arrangement, as shown above, saves on time, protects the saw and minimizes back pain at the end of a long day.

—Mark White, Kodiak, Alaska

TRUING A TABLE-SAW BLADE

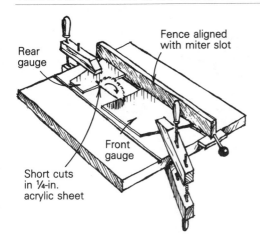

Fence aligned with miter slot

Rear gauge

Front gauge

Short cuts in ¼-in. acrylic sheet

It seemed that no matter how carefully I aligned my sawblade with its fence, they wouldn't end up quite parallel. This can result in cuts that have burned edges, and even worse, the binding action of the blade against the fence can cause the workpiece to kick back. The drawing above shows an easy technique I developed to align the blade with the fence.

First I lined up the edge of my fence with the miter slot to the right of my sawblade and clamped it in place. Next, with the blade in its highest position, I made short cuts in two pieces of ¼-in. acrylic sheet for gauges (hardboard or birch plywood will also work). With the saw unplugged and the fence in the same position, I loosened the bolts that allow the saw's trunnions to be moved relative to the table. A contractor's saw typically has four bolts— two in front and two in back. By leaving one of the forward bolts pretty tight, I can use it as a pivot point. The bolt that I leave tight will depend on whether I need to move the blade clockwise or counter clockwise.

I clamped the forward gauge to the table so that the blade was in its slot, and then lightly tapped the trunnion assembly until the blade fit correctly in the slot of the rear gauge. When the fit is right, the blade turns freely in the slots, and then I tighten the adjustment bolts. The blade should now be parallel with the miter slot in the saw's table.

—Frederic E. Bishop, Farrell, Pa.

RADIAL-ARM SAW STAND

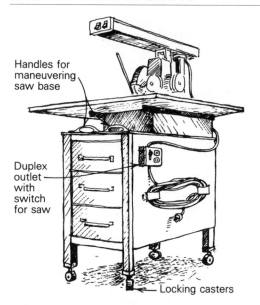

Handles for maneuvering saw base

Duplex outlet with switch for saw

Locking casters

My radial-arm saw came with a rickety bolted-together sheet-metal base. Whenever I moved the saw to a new job site, the base racked and twisted enough to throw the saw out of adjustment, wasting my time and causing me frustration. Clearly, this saw needed a sturdier base.

My search finally led me to a store that sells second-hand office furnishings. There I found old steel desks that consisted of a top that spans a pair of pedestals. Most of the pedestals I found are of robust, welded construction, and they make ideal bases for saws, jointers, shapers and other standing power tools. The pedestals with four legs are the easiest to adapt, requiring only a plywood top and large, locking-type casters on each leg.

My saw base now has a permanently connected 50-ft. electrical lead and side-mounted duplex outlets (ground fault circuit-interrupter type) for plugging in small power tools (drawing above). I added sturdy locks to each drawer so that I can secure blades, bits and hand tools. The unit is stable and rigid, and although heavy, it's easily loaded on and off the truck with a couple of 2x10s as ramps.

—*Karl Juul, Slingerlands, N. Y.*

NON-SKID MITER BOX

While casing windows with some fancy trim, I had problems
keeping the stock from being pulled into the blade. I was cutting
prefinished hemlock casings at a 45° angle for picture-frame joints.
Even though the 80-tooth carbide blade was sharp, it still pulled the
stock inward, leaving the cuts a bit irregular.

To remedy the problem, I used contact cement to glue 120-grit
wet/dry sandpaper to the bed of my power miter saw—one piece
on either side of the blade. It worked very well. The friction of
the sandpaper was just enough to hold the casings fast as I made
my cuts.

—*Jim Esser, Bellingham, Wash.*

BLADE GUIDE

A rolling bandsaw is a good way to cut curves in heavy timbers, but
if the timber is already part of the house, you've got to take the saw
to the wood. We had to remove 3 in. from a 6x10 post to gain room
for a stairway, and we used a Sawzall to make the cut. We guided
the blade on the open side with a wrench, as shown in the drawing
above. This operation takes two people and a fair amount of time,
but it works very well.

—*Jeff Morse, Tomales, Calif.*

JIGSAW BLADE GUARD

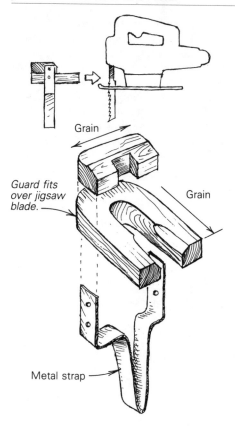

Grain

Guard fits
over jigsaw
blade.

Grain

Metal strap

Being an urban carpenter, I carry a lot of tools up and down stairs. I leave their cases in the truck because they weigh a lot, and pack the tools up to the job in pouches. This won't damage most tools, but I found that the protruding blade of my jigsaw was either breaking off or jabbing holes in the pouch.

I finally got fed up with this and made the blade guard shown in the drawing above. It's basically a little wooden horseshoe with a Y-shaped metal strap (made of flattened copper water line) screwed to it. I beefed up the short-grain portion of the horseshoe with a second layer of pine. Then I whittled away at the horseshoe with a utility knife until I got it to fit tightly between the body of the saw and its base. It's ugly, but it works.

—*Arne Waldstein, New York, N. Y.*

ACUTE ANGLES ON THE CHOP SAW

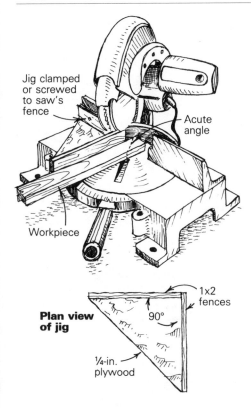

Jig clamped
or screwed
to saw's
fence

Acute
angle

Workpiece

**Plan view
of jig**

1x2
fences

90°

¼-in.
plywood

A couple of years ago I was doing some trim with my friend Marcos Bradley. He was running baseboard around a series of odd angles—angles he couldn't readily cut with his chopsaw. After some thought he assembled a jig similar to the one shown above.

Use clamps or screws to secure one of the jig's fences to the saw's fence. Clamp the workpiece to the jig (block under the far end of long pieces) and you're all set to cut accurate acute angles.

—M. Felix Marti, Monroe, Ore.

ROUND TENONS ON THE TABLE SAW

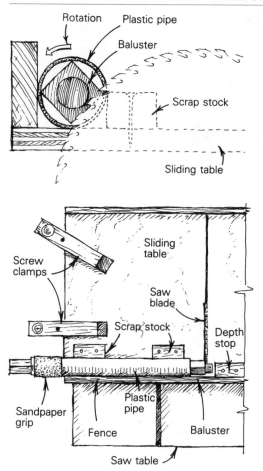

A quick way to make round tenons on square baluster stock is to clamp the sliding table down and screw on some scrap stock to guide a length of plastic pipe toward the sawblade (drawings above). The pipe should be of a diameter that tightly holds the baluster by way of friction.

By adjusting the blade height I can control the diameter of the tenon. To cut the tenon, I simultaneously rotate the plastic pipe the same direction as the sawblade and ease the pipe toward the depth stop.

A thick carbide blade should be used for this operation. And a piece of sandpaper taped to the pipe will provide a sure grip as you turn the pipe. Finding pipe of the right diameter can be tricky. In one case I used a fishing-rod tube and sanded the corners of the baluster stock to achieve the right fit.

—Lauren C. Watson, Newport, N. Y.

HOLE-SAW HELP

Core-type hole saws frequently bind in stock over ¾ in. thick, but even when they do make it all the way through a piece of 2x stock, the core gets lodged inside the saw's cylinder. To avoid this problem, drill partway into the stock and withdraw the saw. Then take a flat-bladed screwdriver and drive it into the plug, with the grain. The plug will split easily, and it can be removed in two or three pieces. Keep cutting through the stock in this manner, and the job will go much faster.

—Norm Rabek, Burnsville, N. C.

FREEING PLUGGED HOLE SAWS

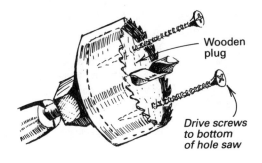

Wooden plug

Drive screws to bottom of hole saw

Many's the time I found myself with a plug of wood securely embedded in my hole-saw chamber. The easiest way to clear up the problem is to run a couple of screws into the plug, as shown in the drawing above. As the screws make contact with the bottom of the saw, they'll begin to back out the plug.

—Ray Brown, Polebridge, Mont.

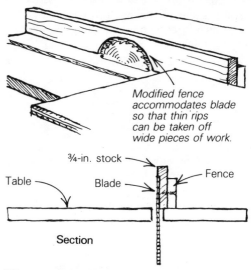

Modified fence accommodates blade so that thin rips can be taken off wide pieces of work.

¾-in. stock

Table — Blade — Fence

Section

Whenever I do finish carpentry on site, I use a small, portable table saw for everything from trim to cabinet work. Unfortunately, the diminutive table limits the width of the work that can be passed by the blade. Sometimes I have to take minuscule rips off the edge of a large piece of plywood, so I devised the modified fence shown in the drawing above to allow close shaves on the little saw.

To modify the fence, I screwed a ¾-in. thick, knot-free piece of wood to the blade side of the fence, making sure that the screws were well away from the path of the sawblade. Then with the blade lowered below the level of the table, I positioned the wood so that it was flush with the outside edge of the blade. I turned the saw on, and slowly raised the blade to full height, cutting a half-moon void in the wood fence.

I use this fence along with an 80-tooth carbide blade to take 64ths off wide stock.

—Jeffrey S. Janssen, Oakland, Calif.

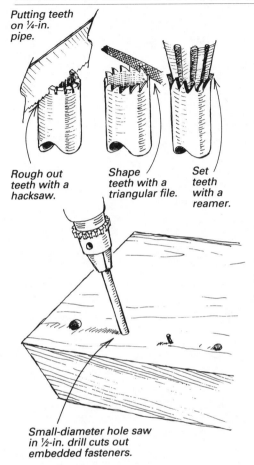

Putting teeth on ¼-in. pipe.

Rough out teeth with a hacksaw.

Shape teeth with a triangular file.

Set teeth with a reamer.

Small-diameter hole saw in ½-in. drill cuts out embedded fasteners.

For those among us who like to recycle wood, the perpetual problem is getting the embedded fasteners out of the wood. Some of the nails and screws have rusted right into the wood fibers, and pulling them is impossible because they've lost their heads long ago. To dig them out without making a mess of the surrounding material, I turn to my homemade hole saw. As shown above, I use a hacksaw and a triangular file to put teeth on the end of a length of ¼-in. steel pipe. Then I use a reamer to give the teeth a little set.

If you don't have a length of ¼-in. steel pipe, you can use copper pipe instead. Better yet, try a push rod. The best hole saw I've made so far started out as a push rod from a diesel engine. I cut the rod in half and put teeth on the hollow center section of the rod. The steel is thin and takes a sharp edge.

—Don Stevenson, Woodland, Wash.

SINGLE-SETUP DRAWER BOXES

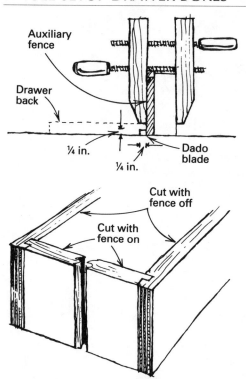

I make my drawer boxes out of ½-in. thick maple plywood joined at the corners with housed rabbet joints. And for the sake of efficiency, I fabricate all the parts on the table saw with a single dado-head setup and an auxiliary fence. The important dimension in setting up the saw to make the cuts is half the thickness of the plywood, ¼ in. in this case. The dado blade is ¼ in. wide. It is set ¼ in. above the saw table, and the fence is positioned ¼ in. away from the dado blade. The auxiliary fence is also ¼ in. thick.

To use the setup, I begin with the auxiliary fence clamped in place. I cut the rabbets on the drawer fronts and backs, using wide pieces of plywood stock that have been cut to length but not to width. Then I remove the auxiliary fence and repeat the procedure for pieces of plywood cut to the correct length for the sides of the drawers. With the dadoes and the rabbets cut into large blanks, I can rip them down to the correct width for the drawers with the rabbets already plowed. This allows me to cut down the number of pieces passed over the dado blade.

By the way, if you set the clamps on the auxiliary fence at the right height, they will act as hold-downs to keep the stock on the saw table as it passes over the dado head.

—Ross Fulmer, Kamuela, Hawaii

TABLE SAW SANDING DISK

Have you ever wished you had a disk sander? If you have a table saw, you've already got one. Simply grind the teeth off an old blade—if you have a 10-in. saw, a 12-in. blade with the teeth removed will likely fit. Stick a sanding disk onto the toothless blade with some contact cement, and you're ready to go. You can even use your miter gauge to assist in sanding different angles.

—Roger A. Bowyer, Dayton, Ohio

5

NAILS AND SCREWS

DRIVING WITHOUT A HAMMER

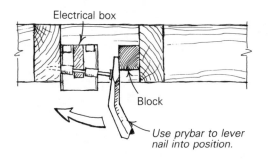

Electrical box

Block

Use prybar to lever nail into position.

More than once I've had to drive a nail in a place that couldn't be reached because a stud or a joist was in the path of the hammerhead. One such situation occurs when studs are closer than typical layout, and an electric box has to be nailed between them. When this happens, I get out out my 2-ft. prybar and lever the nail home. As shown here, I use a block of wood against one of the studs to bring the flat of the prybar perpendicular to the nail. As the nail is embedded, I add shims or a thicker block to keep the flat perpendicular to the nail—otherwise leverage is lost and the nail will go in crooked.

—Steve Sawtelle, Leesburg, Va.

HOLDING LITTLE NAILS

Driving nails with a hammer can be either a satisfying or frustrating experience. One frustration I've learned to avoid is the predicament that occurs when space is tight and the nails are small—like nailing inside the corner of a box. It's easy to hit your fingers or the box, and very difficult to drive the nail straight.

In this situation, I've learned to reach for a strip of cardboard. I push the nail through its edge, right near the end, and use the cardboard to hold the nail over the target. Once the nail is set, I can pull the cardboard free and load up the next nail.

—Steve Mott, San Diego, Calif.

NAIL PAILS

The boxes in which nails come from the lumberyard just don't hold up as I travel from job to job. Even if they don't break, they are awkward to carry, and since framing jobs inevitably require large quantities of different kinds of nails, I need a sturdy container for each type. My solution to this problem is the ubiquitous 5-gal. plastic bucket—cut down to size.

First I remove the handle from each bucket. Then I cut the bucket down to a height of roughly 10 in. I fit the handle into a pair of new holes and then add plywood partitions to isolate different nails. These nail pails are stackable, and their comfortable handles make them easy to carry.

—Jonathan P. Nehemias, Frederick, Md.

NAIL POPS

Often an annoying nail pop on newly hung, taped and painted drywall can be flattened without having to refill and paint it. I take a 6-in. taping knife and flatten it out over the pop. I then give a sharp rap with a hammer to the area of the knife over the pop, and at least half the time, the knife distributes the blow well enough to flatten the spot without marring the paint. A quarter of the time touch-up painting is required, and unfortunately, remaining instances will need recoating and repainting. But the odds are in favor of this technique, and I've never damaged a knife doing this.

—Robert H. Brereton, Minneapolis, Minn.

NAIL PICKUP

Here's a refinement on the old idea of using a magnet on a string to pick up nails and other ferrous-metal debris on a job site. I placed a large horseshoe magnet inside a plastic container (mine originally held ricotta cheese). As shown in the drawing above, I attached a string to the top of the magnet and fed it through a hole in the center of the lid. Then I put a couple of strips of duct tape around the edge of the lid to keep it attached to the container. The result: a magnet that easily picks up metal, yet at the jerk of the string, all bent nails, screws and flashing snippets fall off and land in my recycling receptacle.

—*Raja Abusharr, Eugene, Ore.*

SUBSTITUTE NAIL SET

I've found that a 3-in. powder-actuated fastener makes a good substitute for a nail set. These nails are made to withstand the shock of being driven by a gunpowder charge into concrete, and so far I haven't been able to bend one using it as a nailset. They have a plastic collar on them that makes them easy to hold, and they're cheap enough so that I don't mind if they get misplaced. Before I use one for this purpose, I file its tip flat. And to be on the safe side, I wear eye protection when I'm pounding on it.

—*Ross Elliott, Lanark, Ont.*

INTENTIONALLY BENT

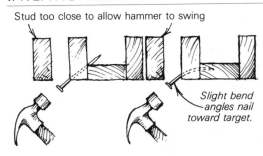

Stud too close to allow hammer to swing

Slight bend
angles nail
toward target.

Here's a simple and effective way of nailing in tight places where your hammer is restricted, and the only angle of drive that you can use will result in the nail not catching both pieces of wood to be attached. Using your hammer, put a slight bend in the nail before driving it. Driven normally, the curvature of the nail will bring it right around into the required area as shown in the drawing above. This method can make the last-minute installation of backing for bathroom fixtures, cabinets or drywall almost bearable.

—Christopher Zane Nestor, Onalaska, Wis.

PULLING NAILS

An elementary but important carpentry technique offers the correct way to extract a "frozen" nail with a claw hammer. With the recalcitrant nail held firmly between the claws, lever the handle down toward the work at right angles to the hammer head.

This method offers greater leverage and control than the more common practice of lifting the handle up and away from the work. The claws have to be sharp and the handle has to be strong— requirements for any good hammer.

—Dick Haward, Nehalem, Ore.

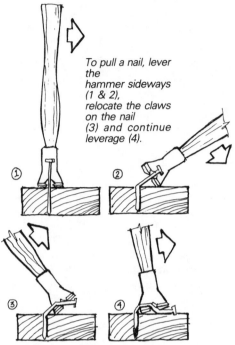

To pull a nail, lever the hammer sideways (1 & 2), relocate the claws on the nail (3) and continue leverage (4).

Dick Haward's tip on pulling nails (see p. 61) is a good one, but it can be carried further. Sometimes the hammer claws will not grip the nail tightly enough to avoid slipping as pressure is applied (steps 1 and 2, drawing above). To finish the job, relocate the hammer claws on opposite sides of the nail (step 3) and re-apply pressure, locking the nail into the claws and lifting it out (step 4). This works especially well in the case of a nail that has "lost its head."

—*Ryan Nevins, Venice, Calif.*

REPAIRING SCREW THREADS

I run across a lot of loose screws, especially in locksets and door hinges. Even in cases that seem hopeless, epoxy glue can be used to provide better purchase for the threads.

First, I comb the screw through my hair to put a thin layer of oil on it. Then I poke some epoxy into the hole, daub some on the screw, and insert the screw into the hole. Once the epoxy sets up, the oil on the screw makes it easy to remove the screw if necessary.

—*Vern Wall. Phoenix, Ariz.*

DRIVING SCREW HOOKS

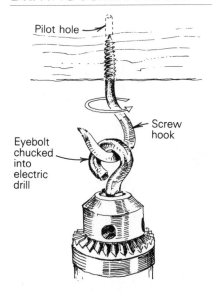

Pilot hole

Screw hook

Eyebolt chucked into electric drill

The next time you have to drive a screw hook into a piece of wood, drill a pilot hole and then chuck an eyebolt into your variable-speed electric drill. Once you've got the screw hook started into the hole, you can slip the eyebolt over the hook, as shown in the drawing above, and use the drill on low rpm to drive home the hook. I've used this method to drive ⅜-in. by 3-in. hooks in just a few seconds.

—John Zouch, Chicago, Ill.

GREASING SCREWS

As a cabinet installer, I drive a lot of screws into hardwoods, and I've found that silicone caulk is the best lubricant. It also acts as an adhesive once the caulk has cured. A so-called "used-up" tube of caulk has enough dregs left in it to grease dozens of screws. So that it doesn't take up too much room in my tool caddy, I cut away the part of the caulk tube that no longer has any caulk in it, and I cut the nozzle off at its base. I use a piece of duct tape to cap the hole when I'm not using the tube.

—Jim Roberts, Charlotte, N. C.

PLUGGING A TOE-SCREW

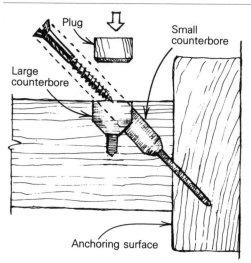

During the course of my work, which is primarily finish carpentry, I find it necessary to angle a screw in toenail fashion from time to time. A typical example is the intersection of a handrail with a wall. The problem is that if the screw is countersunk and counterbored, the angled counterbore can't be easily capped with a plug or a button.

I solve this problem by first drilling a larger counterbore—usually ½ in.—at a right angle to the surface of the stock. I make this hole as shallow as possible. As shown in the drawing above, the second counterbore—usually ⅜ in.—begins inside the larger counterbore and angles toward the anchoring surface.

It takes a little practice to know exactly where to begin the hole for the screw inside the first hole and to drill the screw's counterbore deep enough to allow the screw head to clear the bottom of the plug. So try it on some scrap first.

—*Jeffrey S. Janssen, Nevada City, Calif.*

REMOVING DAMAGED SCREWS

The next time you have to back out a screw that has a stripped head, try using the auto mechanic's method. Apply a dab of valve-grinding compound (available at auto-parts stores) to the tip of your screwdriver. You will be amazed at how well it works to increase the purchase of the screwdriver's tip on a damaged drive slot, no matter what the configuration.

—*Andrew Flowers, Garden Homes, Ill.*

INVISIBLE FASTENERS FOR DECKS

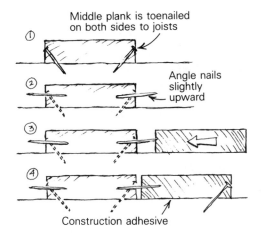

① Middle plank is toenailed on both sides to joists

② Angle nails slightly upward

③

④

Construction adhesive

Some of our customers don't like to look at the exposed nailheads on a typical face-nailed deck. As a result, we've developed the system shown above to secure deck planks to their joists.

With our method, we begin in the middle of the deck with a piece that is toenailed on both sides to the joists. Then we drive 8d galvanized nails halfway into both sides of the plank on 24-in. centers, and cut their heads off with bolt cutters or a pair of beefy sidecutters. We drive subsequent pieces of decking onto the protruding nail shanks by pounding on their sides with a hammer and a block, or by using pipe clamps on pieces that are longer than 10 ft. or so. Then the new plank gets toenailed to the joists, and the side-nailing process begins again. To make sure everything stays put, we also squeeze some construction adhesive between joists and planks. This method leaves no exposed nails to work their way upward, and we've found it to be quicker and cheaper than screwing a deck to its supports.

—Bill Hart, Templeton, Calif.

SPIKED DECK JOINTS

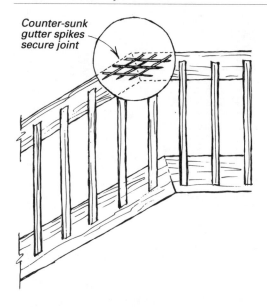

Counter-sunk gutter spikes secure joint

I build a fair number of decks, and every one seems to have a couple of joints that want to open up. The sun and the weather pop out wood plugs over lag bolts, and toenails creep out until their heads stand proud. The drawing above shows how I use aluminum gutter spikes driven in opposing directions to solve problem joints with readily available tools and hardware.

I start by cutting the head off a steel gutter spike with a pair of bolt cutters. The cut end of the steel spike works like the cutting edge of a drill bit, allowing me to use the spike to bore pilot holes for the aluminum spikes, which have a slightly bigger diameter. Holes drilled, I drive home the aluminum spikes, cut off their heads and countersink them with a piece of the steel spike. I fill the resulting hole with a dab of urethane caulk.

The aluminum spikes are barbed and hold tenaciously. They are small enough in diameter for me to use several alongside one another in a staggered pattern, and they are long enough to reach just about any part of a joint.

—*John Trim, Nashville, Tenn.*

PATIENCE AND RECYCLING

To make way for our deck, I had to remove a built-up flower bed that I had framed with pressure-treated lumber. The wood was still in fine shape, and I planned on saving it for future projects. But when I tried to remove the numerous hot-dipped galvanized nails embedded in it, I just snapped off their heads. On the fourth nail I cracked the handle of my framing hammer. By this time I was so disgusted that I chucked the wood in the corner of the garage and went back to the deck project.

A couple of months later I returned to the pile in the garage with the intention of cutting it up for disposal. I decided to try once more to remove the nails, and the first one came out so easily I almost fell over backwards. The lumber had dried out, and in the process had shrunk away from the nails. A little patience works wonders.

—Brad R. Johnson, Chicago, Ill.

6

BITS

SHARPENING ROUNDOVER BITS

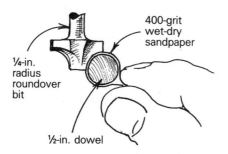

400-grit
wet-dry
sandpaper

¼-in.
radius
roundover
bit

½-in. dowel

I found a quick and easy method for sharpening high-speed steel
router bits when I was doing some ¼-in. radius roundovers. I didn't
have the appropriate stone or file to sharpen the bit, so I took some
400-grit wet/dry sandpaper and wrapped it around a ½-in. dowel.
By pulling the paper taut and holding it between my thumb and
forefinger, as shown in the drawing above, I was able to use the
paper like a file to sharpen the bit quickly. Now I keep an
assortment of short lengths of dowel and 400- to 600-grit paper on
hand to touch up all my concave bits. You can also use sandpaper
wrapped around a wood block as a sharpening-stone substitute for
honing chisels, knives and plane irons.

—*Eric Urani, Adah, Pa.*

DRILL-BIT DEBRIS REMOVAL

While reinstalling some oak moldings on a restoration job, I experienced an annoying problem. The drill bit I was using repeatedly clogged with a mixture of oak shavings and the old finish that coated the moldings. Its flutes were so tightly packed with the hardened mixture that it was impossible to clean the bit by hand.

My solution was to dunk the bit in a cup of water. Upon contact with the water, the stuff clogging the bit soon expanded and popped free as the water simultaneously cooled the bit. To eliminate electrical hazard, I filled the cup with just enough water to cover the bit flutes.

—Paul L. Crovella, Amherst, N. Y.

SHORTENING DRILL BITS

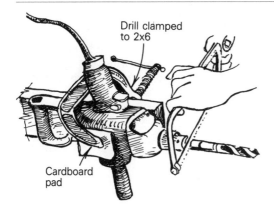

Drill clamped to 2x6

Cardboard pad

Recently I needed to shorten the shank of a ¾-in. hardened steel drill bit. Depressed at the thought of all that time and elbow grease, I devised the method shown above. With the drill securely clamped to a 2x6, I adjusted it to its highest speed, pressed the lock-on button and let gravity and the hacksaw do all the work. A few drops of oil made for a cool cut.

—Dorothy Ainsworth, Ashland, Ore.

DRILL-BIT COUNTERSINK

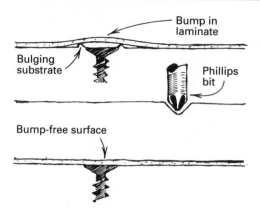

Like most folks, I use self-tapping drywall-type screws for attaching plywood countertop substrate to the cabinets. They go in fast and hold tight, but they also lift up a bit of the wood around the screw's head as it goes in. As shown above, the bulging substrate telegraphs through the laminate surface.

Ideally, each screw should be countersunk with a bit dedicated to that purpose. But that requires a second drill, and yes, more time. The drawing shows a good compromise. I use a spinning Phillips bit to excavate a divot in the plywood before I run the screw home. The divot isn't pretty, but it takes out enough material to keep the fibers from lifting above the plane of the plywood surface.

—*Terence Walker, Seattle, Wash.*

SCREW-BIT ENCORE

To get more life out of drywall screw bits, file the tip when the sides become rounded over. This allows the bit to enter the screw head a little deeper, thus engaging a part of the bit that still has crisp edges.

—*Mick Cappelletti, Newcastle, Maine*

ROUTER-BIT STORAGE

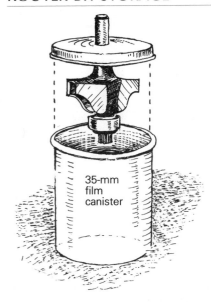

35-mm
film
canister

Good router bits are expensive, and it makes me cringe to think of their finely honed edges bumping together in a toolbox drawer. To protect my bits, I store them individually in plastic 35mm film canisters. As shown here, I drill a hole in the lid the same size as the shaft of the router bit. Then I slip the lid over the shaft of the bit, and snap it tight onto the canister. The cutting edges of the bit are protected, and I don't have to worry about storing them in my toolbox. I like to use clear canisters, such as those used for Fuji film, so I can easily identify the bit without a label. These little containers work great for bits under 1⅛-in. diameter. Forstner and multi-spur bits can also be stored in them.

—Bruce Lowell Bigelow, San Francisco, Calif.

SPADE-BIT SCRIBERS

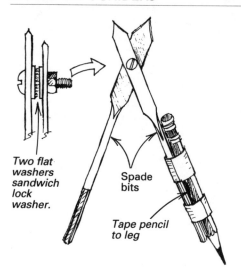

Two flat washers sandwich lock washer.

Spade bits

Tape pencil to leg

I had to scribe a board to fit against an old brick wall last week, but I didn't have my scribing compass with me. However, I did have my drill bits and some assorted hardware. The drawing above shows how I used the parts on hand to make a compass.

I started with a couple of old spade bits, which already had ⁹⁄₃₂-in. holes in them. Then I scrounged a couple of washers, a lock washer and a screw and nut from a junction box. I ran the screw through the holes in the bits, sandwiching the washers between the blades. Voila—tensioned calipers. I taped a pencil to the shorter of the two drill shafts, creating a crude but effective compass that worked just fine.

—*Rob Arthur, Lyndhurst, Ont.*

DRILL-BIT EXTENSION

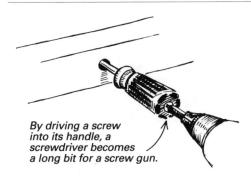

By driving a screw
into its handle, a
screwdriver becomes
a long bit for a screw gun.

When I started to remove a houseful of mini-blind brackets, I
discovered that the bit in my cordless drill wasn't long enough to
reach through the access holes to the screws. Lacking a long
enough bit and the motivation to remove all those screws by hand,
I hit upon this idea: I drilled a pilot hole in the end of my plastic-
handled Phillips screwdriver and ran a screw into it on the highest
torque setting. The screws in the miniblind brackets required a lot
less torque to back them out, so I simply used my cordless drill to
power my hand-held screwdriver, as shown above.

—*Alan Shapiro, Amherst, Mass.*

7

SAWHORSES AND BENCHES

SIMPLE SAWHORSE

With all the fancy substitutes for wooden sawhorses, the real thing is sometimes overlooked. The sawhorse shown above is easy to build and sturdy enough to carry heavy loads. You can pound one together in about 20 minutes, so you don't need to worry about leaving them on the site, unattended at night.

The sawhorse is built of 2x4s, with whatever dimensions suit you best. The two pieces assembled as an inverted T give it rigidity, allow the legs to flare equally and provide a sturdy nailing surface.

—Swami Govind Marco, Montreal, Que.

SAWHORSE ANCHORS

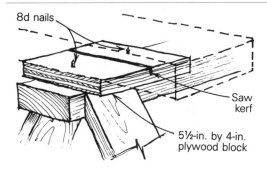

8d nails

Saw
kerf

5½-in. by 4-in.
plywood block

When I have to rip long pieces of 2x framing lumber with a circular saw, I use a pair of ¾-in. plywood blocks to help me secure the work. The blocks are the same width as the lumber being cut. For instance, the 5½-in. block shown in the drawing above is for a 2x6. I tack each block to a sawhorse with two 8d nails, leaving the head of the nail about ¼ in. above the block. Then I snip each nail head off with a pair of side cutters. This leaves about ⅛ in. of nail shank protruding from the block. These protrusions are sharp enough to penetrate a piece of softwood, and long enough to hold it in place while I make my rip cuts. I set the blade depth so the cut enters the block, but not the sawhorse.

—P. J. Woychick, Boise, Idaho

WIRED SAWHORSE

I've attached a weathertight, four-outlet box to the gussetted end of my sawhorse, as shown in the drawing above. This solves the problem of getting entangled in extension cords. Now I only need one cord to serve up to four power tools without having to unplug one, then another one, and another, and so on.

—Paul Dostie, Brunswick, Maine

75

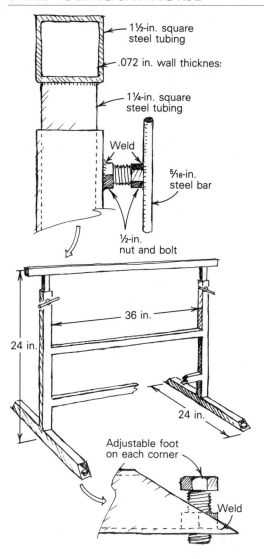

1½-in. square steel tubing

.072 in. wall thickness

1¼-in. square steel tubing

Weld

⁵⁄₁₆-in. steel bar

½-in. nut and bolt

36 in.

24 in.

24 in.

Adjustable foot on each corner

Weld

The ultimate sawhorse should be light as a feather, strong as reinforced concrete, have infinitely adjustable height and collapse to the size of a Swiss-army knife. My design (shown above) using 1½-in. and 1¼-in. square steel tubing hasn't been refined quite to that point, but it sure beats most of the 2x4 or plywood types.

This horse is fairly light, very strong and will nearly double its collapsed height. The ½-in. bolts bind the sliding cross bar tight enough to support lots of weight. The adjustable foot assembly on all four corners takes the wobble out of the horse on uneven concrete floors and allows a screw or nail to be sunk through the nut to hold the horse in place on a construction-site subfloor. Because they nest together, the horses don't take up much room during transport.

Cut the tubing with a hacksaw, reciprocating saw or with an abrasive blade mounted in a radial-arm saw or power miter saw. If you don't have a welding setup, you can keep the cost of the actual welding at a reasonable level by cutting, deburring and drilling the pieces yourself.

—Brooks Heard, Moscow, Idaho

QUICKSET SAWHORSE

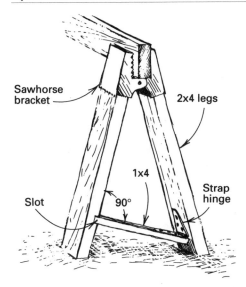

Sawhorse bracket

2x4 legs

1x4

Slot

90°

Strap hinge

Many of the jobs I do are short term—two days or less—and they always require a truckload of tools and materials. The sawhorses, which would be nice to have, are often left at the shop. So I came up with a set of horses that fold up and fit behind the seat of my truck.

As shown in the drawing above, I use 2x4s with metal sawhorse brackets. I screw the brackets to the legs, but not to the rail, so that the rail can be easily removed. At the bottom of one leg in each set of legs I attach a 1x4 with a hinge. Then I fold the 1x4 down until it is perpendicular to the opposite leg, where a slot receives the 1x4. This forces the bottom of the legs apart and causes the top of the sawhorse bracket to bite into the rail. This assembly can be set up and taken down in seconds, and it's surprisingly strong.

—Dennis L. Collard, Colorado Springs, Colo.

SAWHORSE JIG

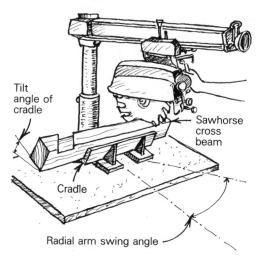

Tilt angle of cradle

Sawhorse cross beam

Cradle

Radial arm swing angle

There's a guy in my neighborhood who builds wooden mason's trestles similar to the sawhorses featured in Tom Law's article in *FHB* #43. To speed up the cutting of the gains that house the legs, he uses a wobble dado mounted on a radial-arm saw, along with the cradle setup shown in the drawing above. The cradle fits on the table of a radial-arm saw, and its tilt determines the splay of the legs in one direction while the swing of the saw determines the splay in the other direction.

—Scott McBride, Irvington, N. Y.

MUD-BUCKET BENCH

Whenever I used to do finish work, I was always looking for a short stool, a concrete block or a bucket to stand on to reach a few pieces of trim. No more. I solved the problem by carrying some of my supplies in a couple of old joint-compound buckets. On site, I empty the buckets and screw through their tops into a 2x10 plank that I bring to the job. The assembly makes a sturdy bench that breaks down easily for transport.

—Curtis Batten, Ellicott City, Md.

FOLDING CHOPSAW BENCH

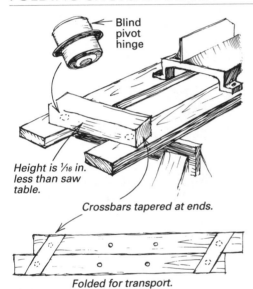

Blind pivot hinge

Height is ⅟₁₆ in. less than saw table.

Crossbars tapered at ends.

Folded for transport.

Here's a design for a folding stand that I made for my power miter box. I started with a couple of 2x boards that I draped across a pair of sawhorses (drawing above). I centered the saw on the boards and drilled holes for bolts that anchor the saw to the bench. Then I cut a pair of 4x4s to the right length for crossbars and ripped them to a thickness ⅟₁₆ in. shy of the height of the miter-saw table. I also tapered their ends so that in the folded position the crossbars remain flush with the boards.

I had some blind pivot hinges left over from a job, so I used them to attach the crossbars to the boards. But a countersunk machine screw or a lag bolt would work just as well.

—Roger Willmann, Columbia, Ill.

ONE-LEGGED HELPER

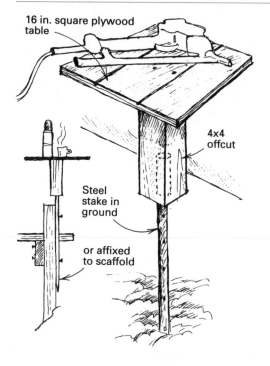

16 in. square plywood table

4x4 offcut

Steel stake in ground

or affixed to scaffold

At the end of a long day, all the typical bending and stooping that it takes to put a tool down and pick it back up can take their toll. On a recent project I solved this ageless dilemma with a one-legged table for my nail gun and fasteners (drawing above).

I began making this shelf with a 4x4 offcut. I drilled a ¾-in. hole in one end, 6 in. deep. To the other end I screwed a 16-in. square piece of textured plywood siding such as T111 (tools slip less on the rough surface). Then I simply pounded a foundation stake into the ground wherever I needed my little table and slipped the 4x4 over the stake. It's held by only one post, so the table is easy to set up on uneven terrain.

By using a longer 4x4, or by attaching the foundation stake to scaffolding, this handy little table rises ever higher to the occasion. And by attaching another square of plywood to the drilled end of the 4x4 for a base, I can put the table to use inside.

—Stephen E. House, Grass Valley, Calif.

QUICK-BUILT T-BENCH

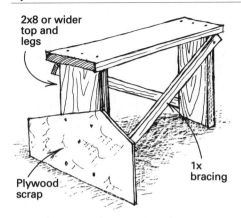

2x8 or wider top and legs

Plywood scrap

1x bracing

A carefully joined sawhorse with splayed legs and compound-angled notches is a thing of beauty; however, the time it takes to make one isn't always available. A similar work platform that is plenty stable and a whole lot faster to build is the T-bench (shown above). To make one, I begin by cutting a couple of legs (16 in. to 24 in. long) out of 2x8 or wider framing lumber. Then I cut a piece about 3 ft. long for the top. Once the top is nailed to the legs, I use a couple of 1xs to brace the legs. Then I stabilize one of the legs with a plywood scrap. Done. Three minutes and one T-bench later, it's back to work.

—*Will Ruttencutter, Chico, Calif.*

8

LAYOUT
AND MEASURING

SQUARE MAKES CIRCLE

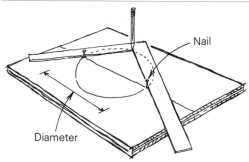

An accurate circle with a diameter less than 14 in. can be drawn quickly with a common framing square. First, draw a straight line longer than your circle's diameter. Mark two points on this line equal to the circle's diameter. Then tack a small nail on the line at each mark, as shown in the drawing above.

Now you can lay a framing square on the workpiece with the inside edges of the square touching the nails. Place a pencil point at the inside corner of the square and slide the square around to each nail. Repeat on the opposite side to complete the circle.

—Matthew C. Jackson, Rapid City, S. Dak.

THE PADD ALTERNATIVE

Can't afford CADD? Try the low-tech alternative: PADD. It's my name for Post-it Aided Design and Drawing. It begins with a trip to the local stationery store, where I can buy Post-it note pads in all sizes and colors with ¼-in. grids printed on them. At this scale, I use the note pads to represent rooms.

As I conceptualize a plan, I can scoot the rooms around on a grid sheet and then stick them down so that they don't move when I put tracing vellum over them, or when the guy next to me opens a window. I draw details such as furniture on them, and I use sheets of different colors to represent separate functions, such as public and private spaces. When I'm really feeling brave, I get out a pair of scissors and start cutting the sheets into rooms with curved walls.

—Paul Chandler, Eugene, Ore.

PENCIL POINT

The ultimate carpenter's pencil is the mechanical pencil—the kind used by people who make working drawings. It has five advantages: The length never changes, so it never disappears in a carpenter's apron; being metal, it's almost unbreakable; the hardness of its leads can be changed to get thin lines for finish work and thick lines for framing; you don't have to rely on your chisel or utility knife to sharpen it; and the lead can be extended to mark precise holes through thick material.

I've had my mechanical pencil in my tool belt for five years now, while watching my fellow carpenters go through boxes of carpenter's pencils.

—Mick Oliver, Ventura, Calif.

ERASABLE SCRATCH PAD

If you glue a small piece of vinyl siding to your tape measure, you'll always have a handy note pad for jotting down measurements. Once you're done with the them, the numbers can be easily wiped off.

—Craig Horstmeier, Davis, Ill.

MORE PENCIL POINTS

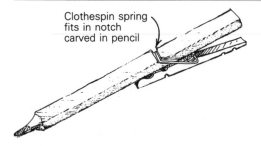

Clothespin spring fits in notch carved in pencil

Although my carpenter's belt has nice pockets for pencils, my instinct is to stick my carpenter's pencil in my shirt pocket. By mating my pencil with a clothes pin, as shown in the drawing above, I devised a pencil that stays in my pocket even when I'm crawling around under a house. Cut the notch with a file or a knife.

—*David Halé, Plainfield, Vt.*

Tom Meehan, in his article "Tiling a Backsplash" *(FHB* #68, pp. 67-69), mentions that his wetsaw kept obliterating his pen-mark cutlines. China markers, or grease pencils, work well under these conditions. They make a pretty wide line, but the lines won't wash off without some scrubbing. You can get grease pencils at office or art supply stores.

—*Gene Swanson, Minneapolis, Minn.*

Working in central Oregon during the winter, I've come to rely on the "Noblot Ink Pencil" to mark frozen boards. Under cold conditions, ordinary pencils leave faint lines on icy wood; the Noblot leaves a dark, inky line. The pencils work equally well on wood that is wet, or dry for that matter. Noblots are made by Faber Castell, and you can pick them up at stationery or art supply stores.

—*Chris Janowski, Bend, Ore.*

BACKUP MEASUREMENTS

A technique I use as a backup when I'm laying out a series of measurements is to mark off critical points (complete with notes) on a length of drywall tape. This makes a useful doublecheck when I'm back at the shop working on a stairway or a load of cabinets.

—*C. R. Kennedy, Rushville, Ohio*

PENCIL EXTENSION

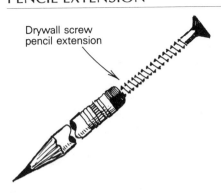

Drywall screw pencil extension

I like to use garden-variety pencils when I strap on my nail apron. Problem is, they eventually become so short that they're difficult to hold, and hard to retrieve from a pocket. To extend the life of a pencil, I run a drywall screw into its typically worn-down eraser, as shown above.

—William R. Roslansky, Woods Hole, Mass.

FRAMER'S LAYOUT TOOL

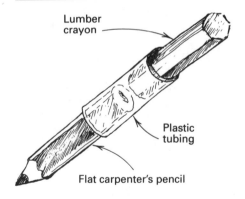

Lumber crayon

Plastic tubing

Flat carpenter's pencil

The writing instrument shown in the drawing above eliminates fumbling around in your tool pouch for a pencil or a lumber crayon and always coming up with the wrong one. It also lets you get more use out of your lumber pencil because instead of tossing your pencil when it gets down below 2 in. long, the plastic tubing lets you use it right down to the nib.

—Todd Sauls, Carpinteria, Calif.

MEASURING ACCURACY

This idea is as old as the Pharaohs, but it's worth repeating. When laying out a measurement that exceeds the length of your folding rule or tape, always lay off the short segment first. For example, if you're using a 6-ft. rule and the measurement is 6 ft. 2 in., lay off the 2 in., then the 6 ft. If you lay off the 6 ft. first, you'll end up with two marks just 2 in. apart—one on layout, and one not. It's easy to forget which one to cut on.

—C. R. Kennedy, Rushville, Ohio

PROTECTION FOR WORKING DRAWINGS

Before we send our working drawings to a job site, we cover them with Con-Tact Clear Covering (Rubbermaid, Inc., 1147 Akron Road, Wooster, Ohio 44691-2596; 216-264-6464). This self-sticking plastic film comes in 18-in. wide rolls and is typically used for book covers, art projects and the repair of torn pages. We've found that prints covered with the product are not only protected from dirt and moisture, but they are also easier to roll and handle. An added bonus is that you can draw on the covering with a marker to elaborate on some detail in the print.

—David E. King, Little Meadows, Pa.

ARCH LAYOUT

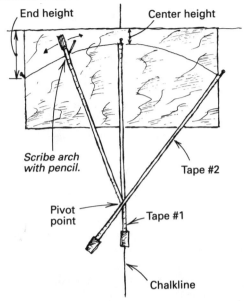

End height

Center height

Scribe arch with pencil.

Tape #2

Pivot point

Tape #1

Chalkline

Here's a way to lay out arches that need to have specific heights at the ends or center, even though the spans vary. First, snap a long chalkline on the floor (drawing above). Lay your material at one end, square to the line. After determining the span, end height and center height, put the end of tape measure #1 at the center height and stretch it out next to the chalkline. Hook tape measure #2 on a nail located at the end height and stretch it out at an angle so that it intersects tape #1. Now move tape #2 until the numbers on both tapes match at the chalkline. Put a nail in the chalkline at this point. This is your pivot point. Hook one of your tapes over it and use it as a giant compass to trace your arch. On huge arches you can make one half and use it as a pattern.

—Spencer Thompson, Santa Monica, Calif.

BALUSTER LAYOUT

Here's a fast way to lay out railing balusters. Get a length of the kind of elastic that is used for waistbands and stretch it so that it's fairly taut. Mark your center spacing on the elastic. Now attach the elastic to one end of the railing run and stretch it out. Move the elastic back and forth until you've got it just right. Mark your centers, roll up the elastic and head for the next railing.

—Tony Scissons, Meadow Lake, Sask.

SCRIBING FOR RECESSED LIGHTS

I had to install some cabinet-grade plywood over a pair of recessed light fixtures, and I wanted to make sure the holes ended up in exactly the right spots. I did it by first skim-coating the backside of the plywood with some water-based putty (joint compound would also work). Then I lifted the piece into position against the light canisters and gave it a push to record their outlines. Perfect circles were impressed in the putty. Then it was easy to find the centers of the circles, scrape away the putty and make my cuts.

—Barney Potratz, Fairfield, Iowa

CARPENTER'S NUMBER CODE

Having followed the debate about English vs. metric measurements, I would like to add the following remarks. My partners and I do cabinets and interior trim work, and we measure everything to $\frac{1}{64}$ in. We not only measure quickly, but we also yell these measurements to a cut man who in turn can quickly mark and cut the stock (he uses a sharp 5h pencil). Our system is so ridiculously simple that I'm surprised it took so long for the necessary language to evolve. We've finally solved the problem of the varying hair thickness. Here's how it works.

In our language $36\frac{3}{4}$ in. is 36-twelve (our code for $\frac{12}{16}$, or $\frac{3}{4}$). In our language $36\frac{25}{32}$ in. is 36-twelve and a half. Halfway between 36-twelve and 36-twelve and a half is 36-twelve strong ($36\frac{49}{64}$ in.). To us, $36\frac{51}{64}$ in. is 36-thirteen shy.

If someone were to hand me a tape with marks every 64th of an inch, I would be lost. But I have no problem splitting a 16th in half and then splitting that half in half. If I have to add a column of numbers using this method, I signify the strongs with a plus sign, and the shys with a minus sign. Then before I add up the numbers, I cancel out as many of the shys and strongs as possible and adjust the total up or down according to how many shys or strongs are left over. To reduce my column of 16ths to whole numbers, I invoke the framers litany—16, 32, 48, 64, 80, 96—while extending one finger for each chant. For example, let's say I've got 87-and-a-half 16ths in my fraction column. I extend a finger for each 16. When I get to 80, I've got five fingers extended. So my answer is 5-seven and a half.

At this point I probably sound like J. R. Tolkien selling used cars. But give this system a chance, and you'll discover its benefits. Just take a small tape and try finding $23\frac{43}{64}$ in. Then take a 25-footer and find 23-eleven shy. Which is easier?

—Jim Chestnut, Fairfield, Conn.

9

BOXES, BELTS, BAGS, BUCKETS AND STORAGE

MAGNETIC NAIL POUCH

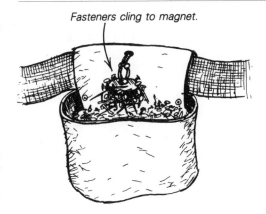

Fasteners cling to magnet.

On a large drywall project you can wear out a pair of gloves (or your fingertips) just by picking screws or nails out of your nail pouch. To prevent this problem, I've attached a magnet to my tool belt with a strong rubber band. I placed the magnet over the center of my nail pouch, as shown above. Located here, the magnet can be extended down into the pouch, where it picks up a supply of fasteners. Then I can quickly pluck them off the magnet without jamming the tip of one under my fingernail.

—*Bob VonDrachek, Missoula, Mont.*

NAILBAG SUPPORT

I've solved the problem of heavy framing aprons by using army-surplus nylon pack suspenders along with the surplus pistol belt. They last forever, and can accommodate most leather nail bags, loops, hoops and tape holders.

—John Friedricks, Lexington, Va.

CARD-CARRYING TOOL BELT

If you don't store your 25-ft. tape measure in the holder provided on your tool belt, consider it as a place to carry 3x5 cards. I can fit about 40 cards, folded in half, in the one on my belt. I'm constantly using the cards for lists and detail drawings, and it makes the formerly neglected pocket a useful adjunct to my tool bag.

—Bob Jewell, Kauai, Hawaii

BELT-BUCKLE UPGRADE

My carpenter friends and I have all agreed for some time that the weak link on a tool belt is the clasp. Some twist-grip buckles are flimsy, and they can all be clumsy to disengage, especially if you have to remove your belt in a precarious spot.

As a remedy, I removed my old clasp and replaced it with a seat-belt buckle. I tried both the push-button type and the spring type, and decided I like the spring type better. Now I can easily remove the belt when I want to, and with one hand at that.

—Evan Disinger, Lemon Grove, Calif.

ELECTRICIAN'S CARRYALL

An electrician has to carry a bewildering array of fasteners and other tiny components, and keeping them organized and close at hand can be frustrating. For me, the perfect way to carry these supplies is with a fly-fisherman's vest. Mine is a multi-pocketed affair that has touch-fastener pocket flaps. It has served me well for years, and it is comfortable enough to let me work in crawl spaces and on ladders.

—Joe Mullane, Redwood City, Calif.

CARPENTER'S CADDY

Through-bolting coffee cans to a plywood backing creates ganged compartments for nails, screws and other job-site odds and ends. Use ¾-in. plywood for the frame, and saw out a handle you can get a good grip on—the kit can get heavy when all the cans are full.

—*Allen Holder, Athens, Ga.*

EARPLUG STORAGE

I use a chainsaw to make a living, building scribe-fit log houses in southern Idaho. Some type of ear protection is a must in this business, and I use the soft foam earplugs that resemble long, miniature marshmallows. The trouble is, when I take them out and drop them into my tool pouch, they quickly become too grungy to re-use.

I solved this problem by putting a plastic 35mm film canister on my tool belt to act as a storage jar. I cut a slit in the pry-off lid, slipped a length of knotted leather thong through the cut, then tied off the thong on my tool belt. Now I've got a handy, clean place to keep my earplugs.

—*Tom Balben, Teton Village, Wyo.*

NAIL-POUCH ORGANIZERS

I've found that the bottom parts of plastic quart-size motor-oil containers make great nail-pouch organizers. Their rectangular shapes fit well into most of my nail bags, and I can use the different colors of bottles to easily identify different fasteners. After draining most of the residual oil from an empty bottle, I remove its top with a utility knife. I use an absorbent rag to swab out the remaining oil and then treat my tools to an oil rubdown.

—*Dan Shumate, Bristol, Va.*

NAIL-BOX HANDLE

When the new boxes of nails arrive on the job site, don't be in a hurry to cut away the plastic straps encircling the box. Instead, leave the straps and use your utility knife to cut away an opening through the top of the box. Now you can use the straps as a handle, and they help to hold the box together until the nails are gone.

—*James A. Brovelli, Studio City, Calif.*

BASKET CASE

Plastic basket for parts and tools affixed to paint shelf.

Know what gets me down? Every time I put something on top of a ladder, I forget about it. Then in the heat of the moment, I move the ladder and off tumbles everything I'm using—hammer, channel-locks, copper fittings and the nastiest of all: open cans of PVC cement.

Not long ago I found the solution to this problem. As shown in the drawing above, I've affixed a 7-in. by 10-in. plastic basket for tools and parts to the bars of the fold-down paint shelf at the top of my ladder. The ones I use, called Stak-N-Tote, cost about $1 at my local hardware store. I use a couple of self-tapping screws to mount the basket. That way I can replace a worn-out basket in a matter of seconds.

—*Paul Penfield, Willoughby, Ohio*

LADDER SACK

Top pocket modified to carry drill

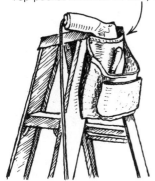

When I climb up a ladder to work, I never have enough room for all my tools and accessories. To solve this problem, I screwed a leather tool pouch on the side of my ladder, as shown above. I used one with a high inside pocket and cut out its bottom seam, which turned it into a drill holster. Now my ladder does double duty by carrying me and my tools.

—*Will Milne, San Francisco, Calif.*

SMALL-TOOL AND MATERIALS STORAGE

When my collection of small tools and oddments got out of control, I knew I would have to find a better storage system. Most small cabinets with many drawers were too expensive, so I sought another solution. A dentist friend sold me his old wooden dental chest, and now I have the perfect storage unit, with drawers that vary in depth from 1 in. to 8 in. Its chest-height marble top is perfect for laying out tools, and by adding casters to the base, I can easily move the cabinet. If you need a similar cabinet, ask your dentist where used dental equipment can be purchased—you might find a bargain.

—*Keith Ojala, Birmingham, Mich.*

ON-SITE LUMBER RACK

Individual bents made from 1x4s

Flat 1x4 extends the length of rack

Diagonal 1x4 stabilizes bents

Keeping track of the trim stock on a complicated job can be a real headache. If there are multiple profiles and widths of casings, moldings, crowns and base, the space required to store and inventory them properly can be considerable. The drawing above shows how my colleague Mathew Marzynski and I solved the problem on a recent job.

The rack is composed of four individual bents made of 1x4 pine. Each bent is 5 ft. tall, with a 3-ft. crossbar at the top. The two lower crossbars are 4 ft. long. A diagonal 1x4 runs the entire length of the rack, adding stability to the assembly. Another 1x4, run flat along the underside of the top crossbars, further braces the bents horizontally.

We easily stacked 500 bd. ft. of oak on this rack last summer. Excluding the floor beneath the rack, there are seven distinct areas for storing materials, and the rack may be loaded from either the sides or the ends.

—*M. Felix Marti, Ridgway, Colo.*

WORK TREE

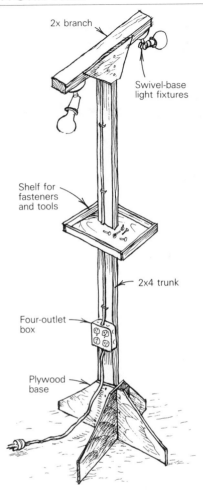

2x branch

Swivel-base
light fixtures

Shelf for
fasteners
and tools

2x4 trunk

Four-outlet
box

Plywood
base

I call the assembly shown here a work tree, and it makes my on-the-job work stations less cluttered and better lit. The structure consists of a base made of criss-crossing plywood legs, a 2x4 trunk and a 2x branch at the top for light fixtures (I prefer exterior-grade fixtures that allow the lamps to swivel). Near the base of the trunk I mount a four-outlet box with an extension cord. Midway up the trunk I have a shelf for tools and fasteners. You can add lots more shelf area to one of these, depending on your taste for luxury versus mobility. This thing is great when you're working in cramped, dark bathrooms. You can also put it next to the sawhorses just to relieve cord and tool clutter.

—Sam Yoder, Cambridge, Mass.

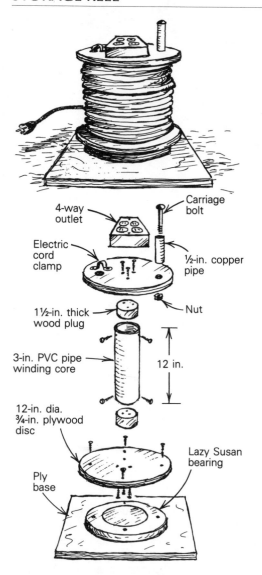

The first thing I've always done when I arrive on a site is roll out my cords from the meter loop to the day's work area. Next, I roll out my nail guns from the compressor to the areas where they will be used. At the end of the day, the last thing I do is roll these same lines up and place them in my truck.

Up until about ten years ago, this was always a time-consuming task. If I didn't make the effort to roll or braid these items neatly at pick-up, the next day's roll-out would be a tangle of 10-ga. cords and ⅜-in. hoses. Then one day I met a fellow with an outfit similar to the one shown in the drawing on the facing page. It took him only about two minutes to roll and store about 300 feet of cords and an equal length of air lines.

That evening, I made up my own reels using scrap ¾-in. CDX plywood, 3-in. Schedule 40 PVC pipe, and a used lazy-Susan bearing. The pipe winding core is affixed to the plywood discs by way of a couple of plugs that I cut from 2x stock with a hole saw. I countersank the screws that secure the pipe to the plugs to keep the edges of the screw heads from abrading the cords.

On my cord reel, I have a 4-way outlet with 100 ft. of 10-ga. cord attached permanently. I find that this is usually sufficient to get me from the meter loop to my work area. My other extensions are attached end to end and rolled up on top of the permanent cord. The 4-way outlet allows me to run two saws at my main area, and run two additional cords from that site to wherever they may be needed.

My hose reel simply has a hole in the top disc through which I pass the first hose end. This anchors the hose and keeps it from spinning on the core while I roll up 300 ft. of hoses connected end to end.

After nearly daily use, my original reels are still working well. The plywood discs have delaminated a bit and the bearings have become a little sloppy, but that's okay. I wouldn't want this work to get too easy.

—*Ric Winters, Georgetown, Tex.*

GLUE BUCKET

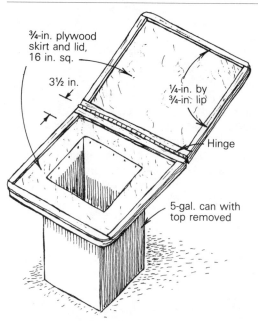

¾-in. plywood
skirt and lid,
16 in. sq.

3½ in.

¼-in. by
¾-in. lip

Hinge

5-gal. can with
top removed

Applying contact cement to laminate work can be a messy
proposition. Here's how I keep it under control. I cut the top off of a
square 5-gal. can (the kind used for some solvents), and attach a
hinged lid and a wood apron, as shown in the drawing above. The
apron stiffens the cut end of the can, and it's important to include
the small strips of wood that act as lips for the lid, otherwise the
inevitable drips will glue the lid shut. Because the lips meet with
such a small surface area, it's easy to pry them apart.

I use a roller to apply the cement because it allows me to spread
it quickly and evenly. The flat inner sides of this glue bucket
provide a surface to roll off the excess glue if necessary. I use a
special roller cover (available from most laminate dealers) that
resists the deteriorating effects of the glue solvents. I find that as
long as the glue in the bucket covers the roller, and it isn't allowed
to dry out, I can keep the roller in the bucket indefinitely.

—Jim Fish, Dale, Tex.

STRAP-ON TOOLBOX

Touch fasteners
at shoulder
strap ends.

Flush handles
at sides

¾-in. by ¾-in.
beveled runners

My toolbox combines ideas from a number of other boxes that I've
seen on job sites. It has flush handles in each end and one on top.
This keeps the surfaces free of obstructions so I can stack things on
it, use it as a work surface and, yes, use it as a stepstool when being
a little taller wouldn't hurt. As shown in the drawing above, there
are beveled runners on the bottom of the box, making it slide
easily on the bed of my truck.

The box is 11½ in. by 12 in. by 32 in., so when it's full of tools, it's
a heavy load. To save my back, I made a loop out of a 2-in. wide,
130-in. long seatbelt-type strap by sewing a 2-in. by 12-in. strip of
touch fastener on each end. I slip the strap through the handles on
the ends of the box so that the joint in the strap is sandwiched over
my shoulder. That way, the touch-fastener bristles are forced
together by the weight of the box. With my toolbox supported on
my shoulder, it's a breeze to open a door, climb a stairway or
negotiate a tight spot.

—Rick Puls, Elkhart Lake, Wis.

HANDSAW BRACKET

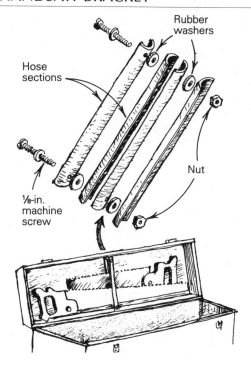

It didn't take me long to split the wood handsaw bracket inside the lid of my toolbox. In searching for a durable replacement, I came up with the bracket shown above. To make it, I first cut a short length of garden hose in half, lengthwise, to give the bracket a low profile. After cutting the hose to length, I drilled holes to match those in the box's lid and assembled the bracket with machine screws and rubber faucet-washer spacers between the hose halves to protect the saw teeth.

—*Sam Yoder, Cambridge, Mass.*

CLAMP HOLSTER

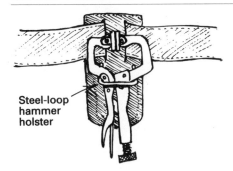

Steel-loop hammer holster

When working with metal studs I add another hammer loop to my tool belt. As shown in the drawing above, the small, locking C-clamps will hang from the loop right where I need them, within arm's reach. The loop I use will hold up to four of the little clamps.

—*W. B. Finke, Kodiak, Alaska*

BELT HOOK

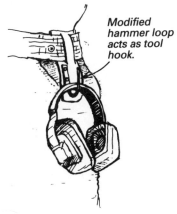

Modified hammer loop acts as tool hook.

This drawing shows how I carry my headphone-type ear protectors when I'm not using them. I modified a steel hammer loop by cutting it in half and bending it upward to make a big hook. It also turns out to be useful for hanging lots of tools, such as my coping saw, my staple gun and my speed square.

—*Eric Roozekrans, Stockbridge, Mass.*

POWER-TOOL BELT CLIP

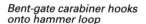

Bent-gate carabiner hooks
onto hammer loop

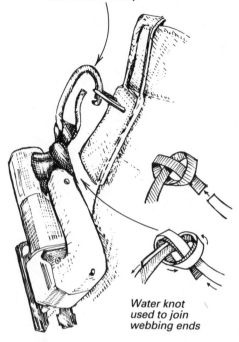

Water knot
used to join
webbing ends

Have you ever wanted to hold a nail gun or a power tool on your tool belt while working? Here's a rig that is easy to assemble, doesn't get in the way and won't accidentally come loose from the belt. And with the rig I can remove a tool from my belt faster than Wyatt Earp could have using just one hand.

As shown above, I use a bent-gate carabiner as a hook to hang on a steel hammer loop. Carabiners are like a link of chain that can be opened by way of a spring-loaded leg. Rock-climbing stores stock them. I thread the carabiner onto a length of 1-in. webbing that is then tied at the ends to make a loop. For this I use a webbing knot that is known variously as a water knot, a ring bend or an overhand bend (detail, bottom right of drawing). To secure the carabiner to the tool, I pass the loop of webbing around the tool's handle and then through itself.

—Cliff Tillotson, Santa Barbara, Calif.

FORM-FIT LEATHER POUCHES

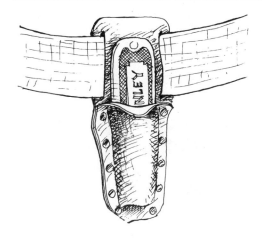

Years ago, I learned from a leather crafter that you can form leather with common rubbing alcohol. I recalled this when I was fixing up my last tool belt. I like to use the old-style fixed-blade utility knife because of its positive, wiggle-free action. But for safety's sake, it needs its own sheath.

My local supplier had a tool sheath that was a little tight for the utility knife, so I soaked the sheath in rubbing alcohol and jammed in the knife. By working the leather with my fingers, I was able to shape the sheath to fit the knife and to roll down the top edge a little to make it easier to return the knife to its slot. Then I set it aside to dry with the knife inside, as shown in the drawing above. Once the alcohol dried, the leather maintained its shape.

Now I've got a sheath that fits my knife perfectly, allowing me to withdraw it easily, while at the same time holding it securely enough to keep it from falling out accidentally.

—Bruce Crooks, Ottsville, Pa.

HAVE NAILER, WILL TRAVEL

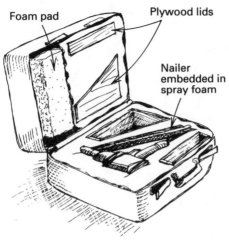

Foam pad

Plywood lids

Nailer embedded in spray foam

I needed a protective case for my framing nailer and its gear, and I found the shell for it down the street at a yard sale. There I picked up an old suitcase of suitable volume. I removed its innards and arranged my nailer and a couple of plywood boxes inside it as shown above. The larger box is for nails, and the smaller one is for accessories. I marked the location of the boxes and the nailer on the suitcase lid and then set them aside.

Next I wrapped the nailer with several layers of thin polyethylene held fast with tape. With small blocks in place to orient the wrapped nailer properly and masking tape along the rim of the case, I spread a layer of triple-expanding insulating foam sealant across the bottom of the case. I placed the nailer and the boxes into the wet foam, pressing them firmly to the bottom. Then I foamed all the remaining voids in the lower half and let the foam cure.

After a couple of days, I used a serrated bread knife to trim away the excess foam and to fine-tune the nailer's recess. Next I placed ¼-in. plywood lids over their respective boxes (separating them by a layer of poly) and coated the lids with a 1-in. thick layer of foam. I gently closed the lid and let it set for a day. After touching up the lid with the knife, I used contact cement to attach a scrap of foam rubber to the top lid to hold the nailer firmly in place. With a modest supply of nails, the case is perfectly balanced.

—*William H. Brennen, Boulder, Colo.*

MUD-BUCKET TOOLBOX

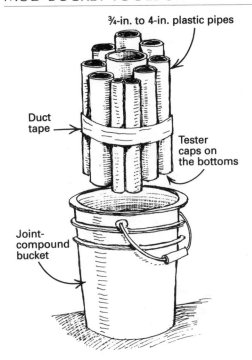

¾-in. to 4-in. plastic pipes

Duct tape

Tester caps on the bottoms

Joint-compound bucket

Joint-compound buckets are just too durable and plentiful to ignore as containers for carrying tools, but I've always been bothered by the way tools end up jumbled together in a bucket. Tools need to be kept in dividers to make sure they're easy to find and to keep any edge tools from getting damaged. I solved the problem by using duct tape to hold a selection of ¾-in. to 4-in. plastic pipes together, as shown above. I put a tester cap on the bottom of each pipe, which allows me to lift the tubes out as a single unit—tools and all—if I need the bucket for some other purpose.

—Mark Russell, Lakeville, Minn.

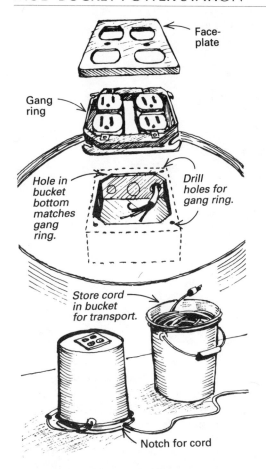

Faceplate

Gang ring

Hole in bucket bottom matches gang ring.

Drill holes for gang ring.

Store cord in bucket for transport.

Notch for cord

Someone was always kicking my extension cord (and its four-outlet gang box) all over the job site. To make it easier for my fellow workers to see the sockets, I attached the gang box to the bottom of a joint-compound bucket, as shown in the drawing above. And with the outlets raised off the floor, I don't have to stoop so far to plug in my tools. The bucket also is a handy place to store my extension cords for transport.

—*George Hennigan, Baldwin, N. Y.*

FLASHING DISPENSER

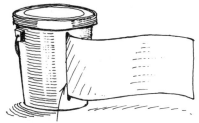

Flashing dispensed through slot

Aluminum flashing is such a pain to unroll that we use the ever-present joint-compound bucket to keep it in line. As shown above, a slot cut in the side of the bucket makes it as easy to peel off a length of flashing as it is to unwind a roll of stamps. The same principle works for large rolls of plastic—just cut a slot in the side of the box and it pulls out like sandwich wrap.

—*Jim Goodrum, Asheville, N. C.*

COPING WITH COIL STOCK

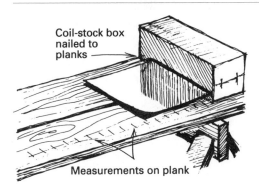

Coil-stock box nailed to planks

Measurements on plank

My helper and I took Jim Goodrum's tip on making a flashing dispenser one step farther. As shown in the drawing above, we remove the coil stock from its box and cut a slot in the box's side. Then we nail the box to a couple of planks draped across a pair of sawhorses. By putting measurements on the plank, we don't need to pull out a tape every time we need to cut a piece to a certain length. We also indicate on the box the amount of stock that we've used. The running total tells us exactly how much stock we have available.

—*Kenneth S. Barnes, Goshen, N. Y.*

CARD-CARRYING CONTAINER

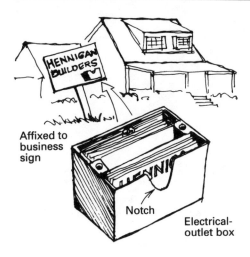

Affixed to business sign

Notch

Electrical-outlet box

I discovered a cheap, practical container for my business cards. As shown here, I took a plastic electrical-outlet box and cut a finger notch in its front. I keep one screwed to the dashboard of my truck for ready access when needed and another one affixed to the sign in front of my job sites for prospective clients.

—George Hennigan, Baldwin, N. Y.

10

TRUCKS AND VANS

TOOLBOX RUNNERS

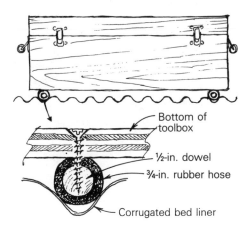

— Bottom of toolbox

½-in. dowel

¾-in. rubber hose

— Corrugated bed liner

The corrugated plastic liner covering my pickup's bed helps to keep the finish in good shape, but its slick surface caused my tool boxes to slide from side to side whenever I turned corners. To make the boxes stay put, I made runners for them out of ½-in. dowels wrapped with ¾-in. rubber hose (drawing above). After drilling pilot holes, I affixed the runners to the bottoms of the boxes with three or four screws each, driven from the inside of the box. Now my boxes can hang on tight on the way to the job.

—*David Gloor, Mountain Home, N. C.*

CARGO PLANKS

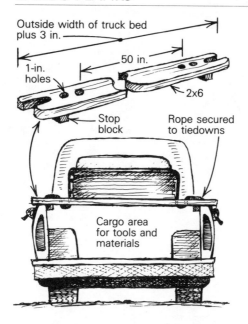

I use a couple of simple carrier boards secured to the top of my pickup's bed to haul panel products while keeping the bed free for tools and other materials. As shown in the drawing above, the 2x6s have stop blocks on each end to keep them from shifting from side to side. Ropes that pass over the payload slip through holes and notches in the 2x6s on their way to tiedowns on the side of the truck bed. The carriers can be made to accommodate oversize widths of materials; they are also strong and easy to store.

—Dan Jensen, Tigard, Ore.

FLAG CONTROL

Wire your red flag to a piece of inner tube cut to resemble a big rubber band. It will readily wrap around almost any projecting load.

—M. Felix Marti, Monroe, Ore.

CARGO KNOTS

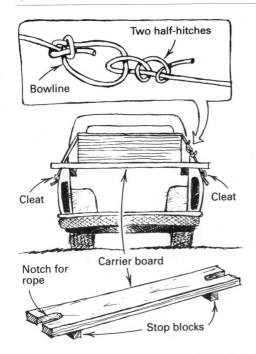

Dan Jensen's cargo platform looks like a handy way to carry sheet goods atop a pickup bed (see facing page), but I prefer my method of tying down the load. I use the trucker's hitch—a bowline and two half-hitches—as shown above. The rope from the left-hand cleat has the loop in it. It extends over the load. Pass the right-hand rope through the loop and haul down on the free end as if your life depends on it. The loop arrangement is akin to a 2:1 mechanical advantage, allowing you to put substantial tension on the rope. Now with your left hand, pinch the loop tightly where the right-hand rope goes through it. This keeps the tension in the ropes while you tie a couple of half-hitches with your right hand.

—*Glenn Bowan, New York, N. Y.*

ROLLING TOOL STORAGE

I needed more storage space in my pickup. I also had a couple of old tool belts that were going unused. To remedy my storage problem, I used some self-tapping screws to affix the old tool belts to the double-walled section of my truck's cab, right behind the seat. Unlike the vinyl "truck organizers" that I've seen, the tool belts have pockets that are designed to hold tools while on the move. And best of all, they're free.

—Bill Thornton, Havertown, Pa.

BEHIND-THE-SHOULDER PLAN HOLDER

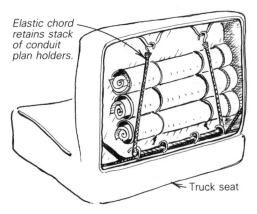

Elastic chord retains stack of conduit plan holders.

Truck seat

Storage space of any kind is a rare commodity in my little pickup truck, so I considered it a windfall to discover a hollow niche behind the seat that's just right for stashing a few sets of plans out of the way. As illustrated above, I stacked three pieces of 3-in. plastic conduit atop one another, and tied them together with tie-wire. On the passenger side, the conduits are capped. I strapped the conduits in place with an elastic cord wrapped around the seat framework, and voilà—no plans rolling around inside the cab.

—Neal Bahrman, Ventura, Calif.

CUSTOM ROOF RACK

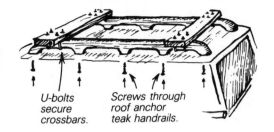

U-bolts secure crossbars.

Screws through roof anchor teak handrails.

I have a sturdy fiberglass shell on the bed of my pickup truck. While it's handy for storing tools and many supplies, I needed a rack on top to carry oversize cargo. I come from a nautical background, so I wanted my new rack to reflect my long-standing interest in finely crafted boats. This led me to the local boat-supply shop, where I found a good selection of teak handrails. They come in various lengths, depending on the number of loops in the handrail. Each loop is about 14 in. long. The rails I selected are the four-loop variety.

I attached the handrails to the roof by running screws with ⅝-in. washers through the fiberglass roof from inside the shell. The crossbars are pressure-treated 2x6s, secured to the handrails with a pair of U-bolts at each connection. Now I've got a sturdy, versatile roof rack that looks sharp, with lots of places to anchor a line.

—Chuck Keller, Marblehead, Mass.

DUMP LINER

This tip is so simple that I'm a little embarrassed to bring it up, but every time I go to the dump I'm reminded that lots of people could put this idea to good use. Before you load up drywall offcuts and old shingles, line your truck bed with one of those blue plastic tarps. Make sure that it's large enough to go up the sides and that it covers the tailgate when you let it down. When the load is light enough, you can use the tarp to pull the load out of the bed. Then you just shake the tarp to clean up. No need to sweep up or pry out the little wood chunks and plaster lumps that get caught in the gap between the tailgate and the bed.

—Nancy Hart Servin, Oakland, Calif.

BACKGROUND NOISE

I do construction work, remodeling mostly, and I like vans because you can lock them, organize tools in them and keep things reasonably dry. I also like my van. We have been together for years now, and I take good care of it.

My van still looks presentable. I repainted it a few years ago, and styles haven't changed much. The brakes and engine have been redone and were giving good, dependable service. Why, then, was I getting an urge to spend money on a new van?

It was the noise. When I drove down a less-than-perfect street— and Washington, D. C., has plenty of those—my van rattled like an old transit bus.

One day while getting a caulk gun from the back of the van, I noticed the support ribs were coming loose from the ceiling and the walls. The adhesive had given up. With silicone caulk and gun in hand, I filled all the seams, reattaching the van's shell and its supports.

What a difference! I can carry on conversations now. And I can hear the radio. In fact, I'm thinking about getting new speakers.

—*Bill Millard, Garrett Park, Md.*

11

HEALTH, SAFETY, COMFORT AND CLEANLINESS

PADDED WORK PANTS

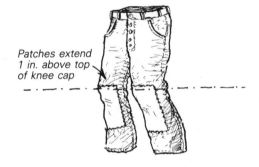

Patches extend 1 in. above top of knee cap

As a general contractor, my clothes take a lot of abuse over the course of a day. To get more use out of a new pair of pants, my wife stitches large padded patches to the knees, as shown in the drawing above. This also makes me much more comfortable when I'm kneeling on cold, damp concrete floors.

She uses bonded batting (used for quilts) as the padding, and twill for the patch. Two 8-in. by 13-in. patches cover padding that measures 7½ in. by 12½ in. She centers the padding on the patch, folds the sides of the patch and straight-stitches perimeter edging. Then she rips out the inseams of the legs to make it easy to sew on the patches. The patches are positioned so that they extend about an inch above the kneecap, and they are attached to the pants with a zigzag stitch.

—Paul Arena, Rome, N. Y.

115

QUICK KNEE PATCHES

Like Paul Arena (see p. 115), I use knee patches to extend the life of my pants, but with one important difference: no sewing is involved. I use a product called Speed Sew, which is available at fabric supply stores. If you can't find that brand, any fabric glue will likely do.

I cut patches from the back of the legs of an old pair of pants, fold the edges over, and glue them down. When the glue is dry, I glue the completed patches to my pants with liberal applications of the glue. I have been using this glue for 15 years (around here they call it "the logger's friend"), and I know it will resist any number of washings or rough wear. Best of all, I don't have to go through the time-consuming sequence of ripping out seams and trying to get a needle through the thick folds of heavy material.

—Patrick Lawson, Sooke, B. C.

WRIST GAITERS

When I have to install fiberglass insulation I make a pair of glove-like shields out of a pair of old gym socks. I cut off the toe and make a little hole for my thumb in each sock. Then I pull each sock all the way up to my elbow, covering all the skin around my shirt cuffs. This keeps the fiberglass from getting into the tender skin on my wrists and arms, while leaving my fingers unhindered to work with utility knives and staple guns.

—William F. Roslansky, Woods Hole, Mass.

LADDER BOOTIES

I was about to set up a stepladder to hang a new dining-room chandelier when I noticed that the ladder's feet were muddy from the last job. It was winter in Chicago—not a convenient time to go outside and hose off the ladder. I thought about spreading a drop cloth and then decided to try something a little simpler. I used heavy rubber bands to secure a shop rag over each foot of the ladder. These socks stayed put no matter where I put the ladder, and I didn't have to worry about snagging the drop cloth or tripping over the folds that inevitably crisscross one.

—Brad R. Johnson, Park Ridge, Ill.

PNEUMATIC DUST MASK

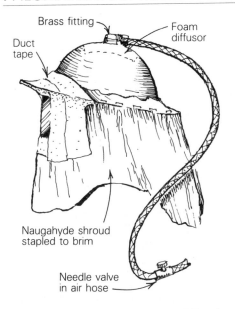

Brass fitting

Foam diffusor

Duct tape

Naugahyde shroud stapled to brim

Needle valve in air hose

Last summer I worked on a log building that required a lot of overhead operations with a router and a belt sander. Consequently, I was often confronted with a suffocating cloud of sawdust and chips raining down on my face. I don't much like particle masks, and goggles are always fogging up, so I tinkered around with an old logger's helmet, some Naugahyde and assorted fittings until I came up with the contraption shown in the drawing above.

My dust mask resembles beekeeper's headgear, with one important exception: it has its own air supply. I drilled a hole in the top of the fiberglass helmet to accept a 3/8-in. brass fitting for an air hose. At about waist level, the hose has a clip that hangs on my belt. Near it, I added a needle valve to control the air pressure from my compressor. Where the air enters the helmet I glued a patch of foam rubber to diffuse the air flow. To the brim I stapled a shroud of Naugahyde that is long enough to hang a few inches below my shoulders. For a window, I used duct tape to attach glass from an old welder's mask.

Although it may look a little funky, this mask works like a charm. In use, I adjust the valve to generate a slightly positive air pressure inside the mask, which keeps out the dust. And as the air flows over the window, it keeps it from fogging up.

—*Dale Baker, Hope, Idaho*

FINDING SPLINTERS

You've got to have a decent pair of tweezers to extract a splinter, but sometimes it's tough to see your target with the naked eye. A loupe can make all the difference in trying to locate a splinter. A loupe is a small magnifying glass that can easily fit into a first-aid kit. Loupes typically come in 5x, 8x and 10x powers and cost about $15. Art supply stores and photography shops are good places to find them.

—Gary Goldsberry, Stillwater, Okla.

CLEANING FOAM OFF YOUR HANDS

I often use commercial spray foam for sealing around pipes, windows and other odd-shaped holes in walls. This kind of small job makes me think that I won't make a mess, so I don't need to wear rubber gloves.

On the rare occasion when I have managed to get the stuff on my hands, I've tried to clean it off with mineral spirits, which are recommended by the foam manufacturers. Mineral spirits work well within about a minute of spraying the foam, but once the foam gets tacky, mineral spirits are less effective. I've found a better solvent to be the carburetor/choke cleaner that you can buy in auto-supply and hardware stores. I find it works well up to 10 minutes after spraying the foam. Of course, all spray foams carry the warning: "Avoid prolonged or repeated exposure to skin," so protective gloves and clothing are the best advice. But for the occasional mishap, when mineral spirits won't work, the carburetor cleaner might be the answer.

—Robert Pauley, Decatur, Ga.

SALAD-OIL SOLVENT

Last summer, while taking a lunch break from staining shingles, I noticed that the oil from my potato chips was dissolving the stain still on my hands after I had washed them with paint thinner and scrubbed them with detergent. Reflecting on how many solvents warn against prolonged skin contact, I decided to pack some olive oil in a squeeze bottle. Using the oil to dilute the stain, followed by a scrub with a handful of sandy loam and some water, left my hands as clean as my partner's paint-thinner-cleaned hands.

—Duff Bogen, Providence, R. I.

LABEL SOLVENT

The next time you need to remove one of those pesky stick-on labels that merchants apply to their goods, spray it with Pam. Sold at the grocery as a cooking spray to keep fried foods from sticking to the griddle, Pam is readily available. You might already have some in the pantry. I've also used it to get the grandkids' bubble-gum out of the carpet.

—M. Kelly Lombardi, Jonesport, Maine

REMOVING LABELS

Anyone who has scrubbed fruitlessly on the gum labels and glued-on instructions that decorate plumbing fixtures will appreciate a can of WD-40. Used with a soft cloth, the lubricant works well as a solvent to lift labels off ceramic fixtures without a scratch. It works equally as well on metal.

—Duane L. Schubauer, Blackhawk, S. Dak.

STAIN REMOVAL

Unfinished wood often gets stained from rusting nails, cement splotches or water leaks. Perhaps you've noticed this kind of staining around the corners of wood-lined skylight wells. To remove such stains, try a solution of oxalic acid and water. Mix ¼ lb. of oxalic-acid crystals with a gallon of warm water and stir until dissolved. Paint the solution onto the stained areas and let it stand for 15 minutes. Rinse the work with cold water and repeat the process if necessary.

Oxalic acid is sold at chemical-supply outlets, and it is poisonous. Be sure to wear rubber gloves and eye protection if you use it.

—Ernie Alé, Santa Ana Heights, Calif.

REMOVABLE SCREEN FLOORS

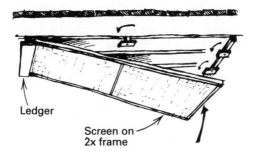

Ledger

Screen on
2x frame

Anybody with a screened porch in a buggy part of the country knows that bugs will get in anywhere they see opportunity—even through the floors. Jerry Germer pointed that out in his article (*FHB* #72, pp. 80-83). But he also noted that his method of stapling screen to the top of the joists made it tough to keep the spaces between the boards free of debris. The drawing above shows the system I use to make removable screens that are affixed to the undersides of the joists.

Each 4-ft. by 8-ft. screen has a 2x2 frame with a spreader in the middle. The 4-ft. dimension is convenient because most decks are framed with joists on either 16-in. or 24-in. centers. The screens are held in place by cleats on one end and on the other by wood blocks that pivot on screws. I try to cover as much of the underside of the deck as possible with this module, but usually there are projections for bracing and other obstructions that require permanent screening around them. That's okay as long as you can reach in from the side to clean the screens.

This isn't an easy or inexpensive way to de-bug a deck floor. But in 10 years of deck building, it's the best solution I've come up with.

—Al Fink, Lynchburg, Va.

CLEANING FOAM OFF THE WALLS

I use foam insulation from an aerosol can when insulating around rough window and door openings. The next day I trim all the excess foam away from the wall and the window frame with a mastic trowel (drawing above). Its serrated edges saw through the foam, leaving the foam flush with the wall.

—Keith Metler, Highland Park, Ill.

SETTING A LADDER

To set an extension ladder at the optimum angle for safe climbing and descending, begin by facing the ladder with your toes against its feet. When your arms are horizontal and fully extended while grasping one of the ladder's rungs, the ladder, no matter how tall, is at the correct angle.

—Derk Akerson, Sacramento, Calif.

BROOMPLOW

2x4 handle

Plywood plow no wider than the narrowest doorway

A sprained ankle can be but a few steps away on a job site that's littered with framing offcuts. That's why I make it a point to pick up construction debris regularly. A push broom, however, can't handle the really heavy stuff. For that I use the plywood plow shown above. It only takes a minute to build one out of scraps. Make the handle about 4 ft. long, and the plow no wider than the narrowest passage or doorway.

—Darrell Ohs, Nanaimo, B. C.

SLICK TIPS

As a carpenter living and working on the Oregon coast, I have always found it a challenge to keep my saws and tape measures clean, dry and free of sand and grit. Sand is hard enough on tools and equipment. Add rain, sawdust, salts and wood sap, and you've got a real abrasive nuisance.

I've found that liberal doses of silicone spray applied to my fully extended tape measures makes them last two to three times longer than usual. Silicone spray coats smoothly, dries cleanly and leaves surfaces slick. Oil-based sprays dry out and leave a greasy film. Spraying the shoe and blade of my circular saw makes for smooth, clean and fast cuts. Ditto for table saws.

—Terry Mackey, Lincoln City, Ore.

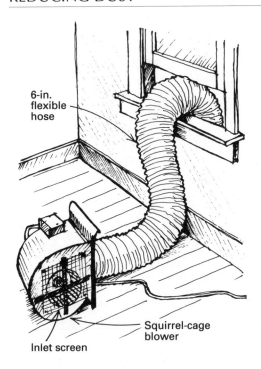

6-in.
flexible
hose

Squirrel-cage
blower

Inlet screen

My work is primarily remodeling, and keeping dust out of inhabited spaces is always a challenge. To that end, I salvaged a squirrel-cage blower out of a central-heating system and attached a 6-in. flexible hose to its outlet port, as shown in the drawing above. This rig allows me to exhaust the dust kicked up by demolition, drywall and carpentry, as well as paint fumes and hot air.

The hose can be snaked through another room to reach a window on the downwind side of the building, and the fan is quiet, so I don't mind running it a lot. The fan is strong enough that it creates a negative pressure in the work area, which keeps construction dust from getting through the inevitable small holes in the plastic that I put between work and living areas. To be on the safe side I put a screen over the blower's inlet. For top performance I use rags to plug the openings around the hose where it exits the window.

—Alan Bellamy, Berkeley, Calif.

SPLICED DEBRIS CHUTE

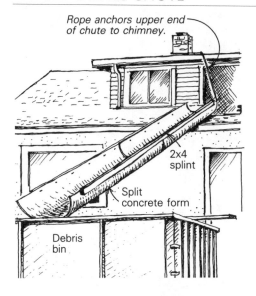

Rope anchors upper end of chute to chimney.

2x4 splint

Split concrete form

Debris bin

We recently did a large residential renovation in which we had to remove nearly every wall on the second floor and a good deal of the roof. Getting this volume of rubbish into the waste bin threatened to take many days and many dollars, to say nothing of our depleting workers' energy and morale. So we decided to make a chute to convey the trash to the bin. We bought a length of concrete form tube 30 in. in diameter and 12 ft. long, and cut it in half lengthwise. As shown above, we overlapped the pieces slightly at the butt joint and then spliced the halves together with 20-ft. 2x4s screwed to the edges of the split forms.

We drilled holes in the top end of the chute for a length of polypropylene rope, which we tied around the nearby chimney. Then we mounted the bottom of the chute to a 2x8 that stretched across the sides of the bin. This arrangement allowed us to place the mouth of the chute near any work area on the second floor or roof. Because we only had to handle the trash once, we were able to keep wood and plaster dust to a minimum. When the rains came, we covered the chute with plastic and kept on using it.

—*Wesley Mulvin and Mary Schendlinger, Vancouver, B. C.*

FILTER FANS

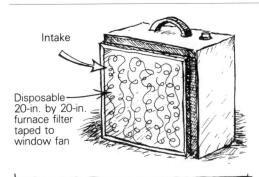

Intake

Disposable 20-in. by 20-in. furnace filter taped to window fan

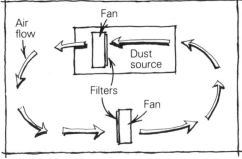

Air flow

Fan

Dust source

Filters

Fan

Plan view

My basement workshop is poorly ventilated, and whenever I do a job that requires lots of sanding, sawing or routing, the air turns hazy with airborne particles. A friend suggested combining a disposable furnace air filter with a standard window fan. Since I had a couple of fans I decided to try it and am pleased with the results.

As shown here, I position one fan near the machine producing the dust and the other across the shop, facing the opposite direction. This sets up an air current that continually directs the dust particles through the filters. I used duct tape to attach the filters to the fans, and I've found that occasionally spraying them with Endust dramatically increases their ability to grab dust particles. To clean the filters, I simply vacuum them when the fans are off.

These filter fans are also handy to take along to the job site during cleanup. A pair of them will clear the air in a good-size room in a couple of minutes.

—*Tom Eckblad, Minneapolis, Minn.*

A QUIETER FAN

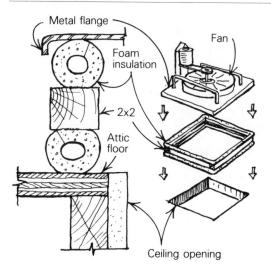

Metal flange

Foam insulation

Fan

2x2

Attic floor

Ceiling opening

In the dead of a hot summer night, when I'm trying to fall asleep, the last thing I want to hear is the whole-house fan in our attic resonating throughout the house. To quiet the thing down, I made the sound-deadening mount shown in the drawing above.

At the heart of the mount is a 2x2 frame, which is sized to match the base of our fan. To the top and bottom of the frame, I glued hollow, 1½-in. O.D. pipe insulation. This is the foamy, pliable pipe wrap used to insulate water supply pipes. I used solvent-based contact cement to affix it to the 2x2s.

I replaced the stock plenum that came with the fan with this homemade frame. The fan simply rests on it. I don't have any scientific data to point to, but I estimate that the foam insulation has cut down the fan's hum and rumble by at least two-thirds.

—Robert M. Vaughan, Roanoke, Va.

BACK SUPPORT

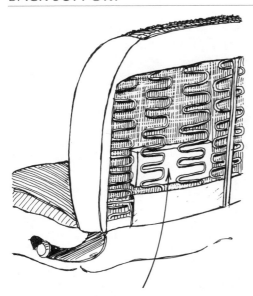

Like many tradesmen, I have a bad lower back. When I bought my used Mazda pickup, I found that the bench seat didn't have enough lumbar support. To change this, I placed a rectangular block of plywood between the springs and the foam back of the seat, as shown above. After some experimentation to find the right size and thickness for the block, I finally settled on a 3-in. by 9-in. rectangle of ½-in. plywood.

—Dave Gent, Ont.

12
SOME SURPRISING IDEAS

REMODELER'S SHIM BOX

As I do remodeling work, I often need small pieces of wood of various thickness to fill gaps. I need them around door and window frames, or when I have to fur out studs or joists to meet existing conditions. To find the right shim, I used to scrounge through debris bins or custom-cut something with the table saw. Now I go to my shim box.

Over the past few years I've been assembling a collection of selected scraps of cedar shingle, Masonite, plastic laminate, plywood and solid wood ranging in thickness from $\frac{1}{16}$ in. to $1\frac{1}{4}$ in. I keep them on edge in a box, grouped by thickness for easy identification and access. Although it looks to some people as if I'm carrying a box of junk onto a job site, I'm certain that the shim box has saved me a lot of time.

—Robert Castle Gay, Seattle, Wash.

TWO-WAY SHIMS

I sometimes use shims in my rough carpentry work, and I make them out of scrap pieces of plywood in assorted thicknesses. I cut them on the table saw so that they measure $3\frac{3}{8}$ in. by $5\frac{3}{8}$ in. At that size, they are ready to shim a 2x4 or a 2x6 without any extra trimming.

—Robert Dziewiontkoski, Greenfield, Wis.

EASY-PEEL MASKING TAPE

Rub well the edges of your masking-tape roll with paraffin, crayons or candle wax and you lessen the chance of paint being pulled up when the masking tape is removed.

—*Robert M. Vaughan, Roanoke, Va.*

REMOVING GLUED-DOWN WOOD

If you need to remove a piece of wood that has been put down with mastic, take a single-strand guitar string and attach each end to wooden handles. Insert the string between the wood and the mastic and pull the handles back and forth. The guitar string works like a cheese cutter.

—*Michael Murphy, Guemes Island, Wash.*

IRIS AT THE CORNERS

When I subdivided some heavily wooded land, I found it almost impossible to keep track of the marker pins in the undergrowth. Following an old-timer's advice, I planted a tuber of iris at each pin. It's the old-fashioned tough-through-any-winter kind of iris that we used to call "flags." Each tuber grew and multiplied, and I'll bet those pins will still be easy to find a hundred years from now.

—*Betsy Race, Euclid, Ohio*

STRING SNARE

The time-honored way to connect stereo speakers or TV antennae is to drill a small hole in the floor, and then run the wires across the basement ceiling. That's fine if you've got a basement, but my house has a 2-ft. high crawl space that is a tapestry of cobwebs and spiders. Rather than crawling 12 ft. to reach the hole in the floor, I put my chimney-cleaning brush to work.

I pushed a length of string through the hole so that it dangled below the floor. Then I attached the brush to a 15-ft. long fiberglass pole and snagged the string in its bristles. After a couple of twists, I had the string firmly entangled. After I pulled it in, I used the string to pull a speaker wire to its final destination.

—*Bruce Knott, Grand Haven, Mich.*

WIRE TWIST

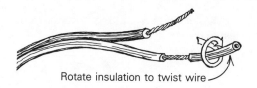

Rotate insulation to twist wire

Multistrand wire has to be tightly twisted before it's wrapped around a terminal or spliced with a wire nut. If you've ever worked with the wire, you know how frustrating it can be.

Getting a good, tight and fast grouping of the strands was a hassle until I came up with the following technique. First, I carefully cut the insulation all the way around the wire. Then as I slowly remove this tube of insulation, I rotate it as shown above.

—Frank R. Melka, West Caldwell, N. J.

EXTRACTING FENCE POSTS

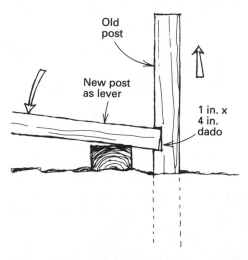

Old post

New post as lever

1 in. x 4 in. dado

When replacing a fence, it is often convenient (and sometimes necessary) to put the new posts in the old holes. The problem, of course, is getting out the old posts. My solution is to lever them out of the ground, as shown above. I make a row of 1-in. deep passes with my circular saw in the side of the post and then knock out the chips to make a dado for the lever. Then I use a post offcut as a fulcrum and a length of post stock for the lever.

—Frank Chester, Victoria, B. C.

IMPROVED POST PULLER

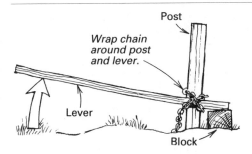

Post

Wrap chain around post and lever.

Lever

Block

Frank Chester's post-extraction method works, but my technique offers an improvement. Instead of notching the old post for a lever, I tie the lever to the post with a chain (drawing above). The chain can be tied to itself with a hook that fits between the links or with a bolt and a wingnut.

The far end of the lever rests on a block, and the lever is lifted instead of depressed to get the post out of the ground. The advantage of this method is that the post can be worked from different positions instead of just one side. A word of caution: protect your back when extracting posts this way. Use your legs, and let the lever do the work

—Thomas Ricci, Lexington, Ky.

MORE POST PULLING

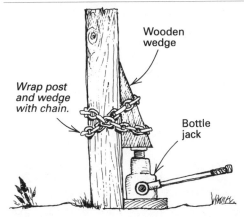

Wooden wedge

Wrap post and wedge with chain.

Bottle jack

Here's yet another way to extract posts. Unlike the techniques of Frank Chester (see p. 130) and Thomas Ricci (above), which require levers and a fair amount of room, my method is very effective in tight spots where you don't have much clearance.

First, cut a wedge from a 4x4 or a 6x6. It should be about 12 in. to 16 in. long. Next, wrap a chain around the post to be pulled and slide the wedge between the chain and the post, as shown in the drawing on p. 131. Lay a 2x scrap on the ground under the wedge to serve as a base for a bottle jack. When you pump that jack's handle and take up the slack in the chain, nothing will keep that post in the ground. When I used to work for the railroad, we raised telegraph poles with this trick.

—Eric Roth, Dorset, Ont.

LAG-BOLT CABINET LEVELER

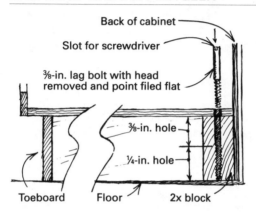

This drawing shows a simple and inexpensive way to level a cabinet when you can't get under it with a shim. It requires nothing more than a 2x block and a ⅜-in. lag bolt. The length of the bolt depends on the height of the toeboard.

To make the leveler, I start by drilling a ⅜-in. hole in the block. The depth of this hole should equal the unthreaded portion of the bolt. Then I drill a ¼-in. hole through the remaining portion of the block. To cut the threads for the bolt, I run the bolt into the hole its full length and then back it out. Now for the modifications.

I lop off the bolt's head with a hacksaw, then use the saw to cut a slot in the bolt's shank. Finally, I cut off the tip of the bolt and file it flat. The bolt is now ready to be installed and run into its hole with a screwdriver. Once it engages the floor, it's out of view below the level of the cabinet's bottom shelf, but it remains accessible.

—Dan Jensen, Tigard, Ore.

POWER-CORD STRAPS

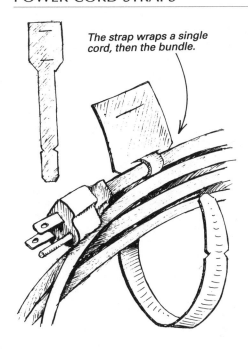

The strap wraps a single cord, then the bundle.

The drawing above shows a strap that I use to keep my extension cords nicely coiled when they're not in use. I make the straps out of defunct inner tubes for truck tires, available for free at my local tire dealer.

To make a strap, I start with a 2-in. wide strip of rubber about 10 in. long. Then I trim the strap so that I've got a 2½-in. long body with a ¾-in. wide tail. I make two slits in the body through which the tail will pass. It's important not to make the slits too wide, or the body will not grip the tail very well.

I fasten the strap on the male end of the cord as shown **above**. Then I wrap it around the coils of the cord and pass the tail of the strap through the remaining slit. The holding power of the strap is improved if I cut two small notches in the tail where it engages the slit.

I use similar straps to keep all manner of hoses in order. I just make them longer to accommodate the increased bulk of the coils.

—*John Schmidt, Sequim, Wash.*

REUSE YOUR SHOES

Unlike shoelaces, touch fasteners such as Velcro aren't worn out by the time the kids outgrow their shoes. So I snip off the fasteners and put them to work again. They make excellent ties for extension cords and bundles of wire. And they can be linked together to make ties of any length.

—*Eugene Capaiu, Rolling Hills, Calif.*

HOPPER PROP

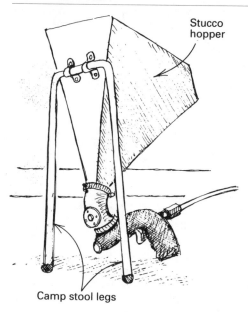

Stucco hopper

Camp stool legs

I've been cussing my spray-stucco hopper for years because it has to be propped up or set in a bucket every time I put it down for a refill or change its position. Then I saw a folding camp stool in my pile of aluminum recycling scraps. I cut the rivets that held the two halves of the stool together and attached one pair of legs to the hopper with pipe hangers and a couple of pop rivets (drawing above). The tight-fitting hangers hold the legs in any position, and now my hopper sits wherever I put it, like a good dog.

—*Dave Echeverria, Corral De Tierra, Calif.*

THE EDUCATED KEY

After 70 years, I still couldn't find my front-door key in the dark. Until now. I sat down and thought about it and came up with the idea shown above. I drilled an off-center hole in the body of the key and put the key back on my key ring. Now I can hold up the ring in the dark, and the educated key will pop out at about a 45° angle. It never forgets.

—*Eugene Capaiu, Rolling Hills, Calif.*

SHAKE AND SHINE

Before I install any pieces of brass hardware on my projects, I toss them into a vibrating case polisher that originally was designed for cleaning cartridge casings. The polisher is a large plastic bowl attached to an electric motor that causes the whole thing to vibrate. The bowl is filled with a dry polishing media. As it vibrates, the fine talcumlike media gets into every nook and cranny—places no cleaning rag will ever go. There are no caustic chemicals to mess with and no elbow grease. Just drop in the hardware and walk away. I bought my case polisher at a local sporting-goods store. There are several different brands and sizes available.

—*Michael Chilquist, Pittsburgh, Pa.*

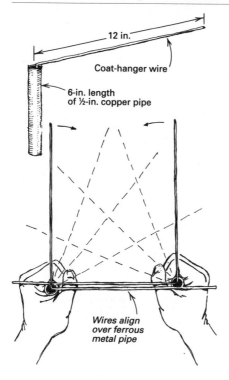

12 in.

Coat-hanger wire

6-in. length
of ½-in. copper pipe

Wires align
over ferrous
metal pipe

There have been many times when I've had to locate underground lines for water, sewer, gas or electricity. After digging too many exploratory holes, my brother showed me an easy way to locate pipes or conduits made of ferrous metals. I was skeptical, but I became a convert.

First, cut yourself two pieces of ½-in. copper pipe about 6 in. long (drawing above). Then take two metal coat hangers and cut them into 18-in. lengths. Straighten the hangers out, and make a 90° bend in each one 6 in. from the ends. You now have what look like two big Ls. Insert the short end of each L into one of the copper pipes.

To use the rigs, hold a pipe in each hand in front of you, about a shoulder width apart. The hangers should move freely in their pipes. I suggest you start with a known underground line so that you can see what to look for. Holding the pipes with the wires pointing forward, slowly walk toward the line. As you approach it, the wires will swing inward. When they cross each other, you're directly over the line. The closer you are to the ground, the greater the accuracy.

—Dennis T. Harbison, West Chester, Pa.

GARDEN BOARDWALK

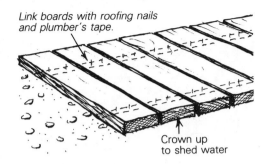

Link boards with roofing nails and plumber's tape.

Crown up
to shed water

I wanted some paths between some of my bushes, but I didn't want the look of concrete stepping stones or gravel. A redwood boardwalk seemed like a good choice, and a nearby lumberyard sells 2x6 redwood culls for almost nothing. I cut the 2x6s into 14-in. treads, and then I linked the treads together with plumber's tape and roofing nails (drawing above). Each segment of my path has between five to nine treads in it, which makes them very manageable, and the treads are arranged with their crowns pointing up to ensure good drainage.

—Victor A. Maletic, Antioch, Calif.

13

QUESTIONS AND ANSWERS FROM BUILDERS

FIXING A BENT SLICK

I recently purchased an old timber-framing slick, which is in excellent shape except for a slight bend in the back. It is made of laminated steel, 3⅜ in. wide. The bend occurs about 7 in. back from the cutting edge, at the point where the laminations stop. I tried clamping the slick in a vise between blocks that were spaced to apply pressure at the point of the bend, but to no avail. I sure don't want to damage the tool. What next?

—Steve Becker, Valatie, N. Y.

Barr Quarton, a custom knife and toolmaker in Hailey, Idaho, replies: It sounds like the slick may have warped while being heat-treated. If the bend isn't too extreme, and if it sweeps away from the flat side of the blade, I would advise leaving well enough alone.

If the bend is a radical one, then the tool will have to be heated completely, straightened, then fully annealed and retempered to avoid setting up more stress in it. This is a complicated process and should be done only by an experienced toolmaker.

INSTALLING JOIST HANGERS

U-shaped joist hangers are a standard item on most framing jobs, but installing them accurately, with the right height and spacing, can be tedious and time-consuming. Can anyone suggest a jig, a fixture, a clamp or a technique that would make the job faster and easier?

—Donald Matthews, Coeur d'Alene, Idaho

Fred Misner of Edmonds, Wash., replies: Because I usually work alone, I developed the following method of installing joist hangers. I nail the hanger flush with, or a little back from, the end of the joist. This ensures a tight fit (otherwise the hanger could hold the joist away from the beam). Next I clamp a short 1x2 or 2x2 stick to the top of the joist with about 1 in. extending past the joist end, using a sliding bar clamp or pipe clamp. Then I set the joist in place with the extending stick resting on top of the beam—this automatically lines up the top edges. Next I hammer home the two barbs on the hanger. When nailing off the hanger to the beam, I angle the nails in opposing directions to reduce the pull-out potential.

If a clamp isn't available or if you prefer a low-tech method, nail the hanger on the joist end, then line things up to your liking and use the barbs on the hangers to rough set the joist. Fine-tune, if necessary, by adjusting the hanger with your hammer until it's in place, then nail it off to the beam.

Todd Sauls of Santa Barbara, Calif., replies: In most cases I attach joist hangers to the beam before installing it. This eliminates the hassle of trying to nail from a ladder or bending over and working from above. First make a T-shaped jig out of scrap pieces of 2x6 and 2x4 (drawing below). The length of the 2x4 should equal the width of the joists. Nail the scraps together with one edge flush, which leaves a lip on the back where the 2x6 overhangs the 2x4. You can

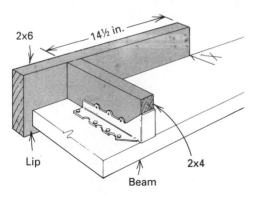

2x6

14½ in.

Lip

2x4

Beam

lay out spacing for joist hangers, or you can cut the 2x6 long
enough to serve as a layout tool. Hook the lip over the top of the
beam, slip the joist hanger over the 2x4 and nail it off.

Fred Mocking of Skokie, Ill., replies: The drawing below shows a jig
that I use to lay out and install joist hangers. It's made with scraps of
joist stock and a length of 1-in. closet rod. On top of each block I
install a 4-in. metal nailing plate with barbs on each end—the kind
used to protect plumbing runs from nails. I screw the plate to the
block, with the plate overhanging the block about ¾ in.

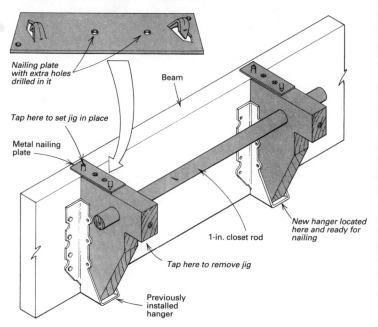

You have to install the first hanger without the jig. Then you slip
the next hanger over one block and insert the other block in the
first hanger. Tapping the barbs into the jim joist or beam will hold
the jig in place while you install the hanger. To remove the jig, tap
up on the block.

Steve Chassereau of Portland, Ore., replies: Here's a jig that not
only aids in the installation of joist hangers but also spaces them
the correct distance apart. To make the jig, use a 2-ft. scrap of
plywood 10 in. or 12 in. wide. In one end of the plywood cut two
slots that will register in a previously installed joist hanger (top
drawing, facing page). Measure for the on-center spacing of the
joists and cut another pair of slots. Slide a joist hanger into these
slots, mark where nails will go and cut notches around the nail

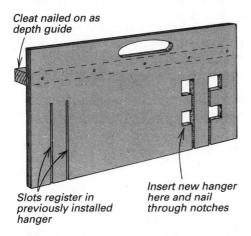

Cleat nailed on as depth guide

Slots register in previously installed hanger

Insert new hanger here and nail through notches

locations. On the back of this plywood jig, nail a cleat spaced up from the bottom edge a distance equal to the width of the joists. This will act as a depth guide. You can even cut a hole in the plywood to serve as a handle.

To use the jig, insert a joist hanger into the second set of slots—spring tension should hold it in place—slide the first set of slots over the previously installed joist hanger and rest the depth guide on the rim joist or beam. Then, nail the new joist hanger in place.

Chris Cartwright of Middlebury, Vt., replies: Because joists can vary in width as much as ¼ in., installing joist hangers before installing joists won't always result in precise framing. I've found the best technique is to toenail the joist in place first, then install the joist hangers.

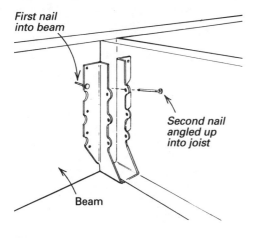

First nail into beam

Second nail angled up into joist

Beam

Snug joist-hanger installation is accomplished by nailing the hanger to the beam first and then securing the hanger to the joist with a second nail (shown on p. 141). Angle the second nail up so that when it is driven home, the joist hanger is drawn up tight against the bottom of the joist. This done, the joist hanger can be nailed up the rest of the way. And yes, all the holes should get a nail. The load-bearing capability of the joist hanger is calculated on the basis of a full complement of approved joist-hanger nails working to resist shear.

THE ACCURACY OF WATER LEVELS

What accuracy can I expect using a water level? When I fill a tube with water (all air removed) and hold the ends next to each other, the water levels are never the same height, and the difference varies every time I move the tubes. Can you tell me the reason for this?

—Rick Strub, Amherst, N. Y.

Tom Law, consulting editor of Fine Homebuilding, *replies:* In its natural state, water infallibly seeks its own level. The problem is that when we put water into hoses or tubes, we're doing something unnatural with it. In order for a water level to work properly, the water has to be able to move freely. The tube cannot be crimped or have air bubbles in it, and the tube has to be open to the atmosphere on both ends (no plugs or fingers blocking either end of the tube). Under those conditions, a water level simply has to work with 100% accuracy.

There are some things that could prevent the level from working properly that you may not think of. First of all, use a tube with ½-in. I. D. or larger. In a small-diameter tube, surface adhesion makes the water's surface concave—the water wants to creep up the inner walls. In a larger tube, there is still the same adhesion, but the surface appears flatter.

If you're using a water level outside in the winter, ice can form and obstruct the flow of water, but you won't be able to see it. The water must be the same throughout. Antifreeze is often used in a water level to avoid freezing (the color of antifreeze also makes the water level easier to read). The antifreeze should be mixed thoroughly with water before the tube is filled. If you pour the antifreeze directly into the tube, it may not mix equally with the water. And the specific gravity of the solution could be greater in one end of the tube than in the other.

WINDING EXTENSION CORDS

I've purchased a 100-ft. 12-ga. all-weather extension cord. Can you tell me the best way to wind it up? There must be many methods used around the country. Are there any manufactured cord winders that are worth using?

—Len Prelesnik, Holland, Mich.

Consulting editor Bob Syvanen replies: To keep my cords in good shape I took to winding them up with the right hand one day, the left hand the next. Even with this system I would periodically have to wind the cord in a coil, flat on the ground, untwisting as I coiled. Now I wind my heavy cords on the ground all the time because it's easy and sure. I don't know that this is the best way, but it works for me.

Cord reels are a neat and quick way to wind cords. The best reels I've seen are those used by video and audio production crews. They are rugged, with bearings and a braking system. They cost from $60 to $200. One source is Markertek Video Supply (145 Ulster Ave., Saugerties, N. Y. 12477; 914-246-3036).

READERS REPLY
Regarding the inquiry from Len Prelesnik (above) about winding extension cords, I use a method that is simple, quick, and makes the cord very easy to unravel.

Hang here.

You double up the cord and lay it out with one half alongside the other. Grasping the plug ends, you begin winding 3-ft. loops with the doubled-up cord, being sure that the plug ends hang below the initial loop to prevent fouling (drawing above). The final 4-ft. length of doubled-up cord is wound around the 3-ft. loops and then poked through top half. This results in a neat cord that stays put and doesn't get tangled, whether it is hung by the convenient loop or thrown into the tool box. To unravel, simply unthread the center hanging loop through the 3-ft. loops, unwrap and cast the entire cord away from you like a life ring on a boat. It will lie flat on the ground without a tangle or a gripe.

—Steve Manning, Webster, N. H.

To avoid the destructive twist of rolling a power cord, I chain it (to use a crocheting term) instead. This is a complex-looking braid pattern, but the cord can be stored and later uncoiled faster than any other method I've found.

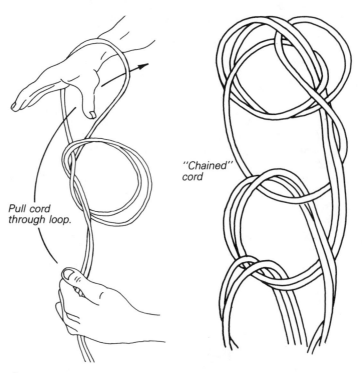

Pull cord through loop.

"Chained" cord

First, double up the cord. Then start at the end opposite the plugs and tie a loose "granny" knot. Reach through the loop in the end and pull out a short length of the doubled-up cord (drawing above, left). This creates another loop. Reach through that loop, grab and pull. Repeat this process (above, right) until you get to the ends of the cord. A typical 100-ft. heavy-gauge cord reduces to a 10-ft. to 12-ft. length. To unbraid, simply pull on the loose end.

—Jim Power, Lawrence, Kan.

I would like to offer a technique that's been used for decades in the television industry. The trick is to alternate the lay of the cord with every wrap. Hold the cord in one hand and form an overhand

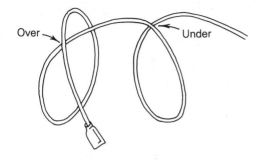

Over

Under

loop, 3 ft. to 5 ft. long, with the other. On the next loop, however, loop the cable underhand (drawing above). Continue alternating until the cable is neatly coiled. The alternating turns will allow the cable to be unwound without twisting.

—*John A. Briggs, Salt Lake City, Utah*

PANEL PILOT BIT

In the "Tips & Techniques" section of *FHB* #49 (October/November 1988), Michael Gornik mentions a router bit that he uses to rout out plywood sheathing for door and window openings. I have dreamt of such a tool and am ecstatic to know it exists, but I can't find one. Can you tell me where I can get this bit?

—*Al Hernandez, Titusville, Fla.*

Charles Miller, a senior editor of Fine Homebuilding, *replies:* The bit you're looking for is available through the mail from McFeely Hardwoods and Lumber (P. O. Box 3, 712 12th St., Lynchburg, Va. 24505; 804-846-2729). In their current catalogue the bit is called a "panel pilot bit with drill point." The cost is $9.91 for the ¼-in. bit and $15.05 for the ½-in. bit.

THE LITTLE NIB ON OLD HANDSAWS

I liked Tom Law's discussion of old handsaws *(FHB* #54 pp. 94-96). I have some old saws myself, one of which has a small cutout and point at the end. Do you know what purpose this served?

—*William R. Phelps, Del Mar, Calif.*

Associate editor Kevin Ireton replies: Although I've heard lots of speculation about the purpose of that little point, I'm inclined to believe the explanation Roy Underhill gives in his book *The Woodwright's Companion* (The University of North Carolina Press, 1983): "The first question many ask about old broad-bladed handsaws is 'What's this little nib for?' They're not on the handsaws that one sees today, but steel handsaws made before the mid-twentieth century commonly had a tiny protrusion, a 'nib,' near the

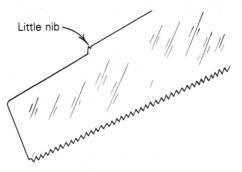

Little nib

toe end of the sawblade (drawing above). You will hear all sorts of explanations—that it's an aid to starting the cut, that it's used for securing a wooden blade guard, or even that it's to keep the saw from sliding out of the carpenter's tool basket.

"Practical men want to see a practical use for all things, but the origin of the nib is essentially a decorative one: it's a vestige of the decorative filigree work found on the early Dutch and Swedish handsaws. It's something akin to the pathetic chrome strip found on the front fenders of Buicks where once-proud Cruiserline Vent-i-ports formerly reigned."

BACKUP POWER

Thanks to Hurricane Bob, last fall we lost electricity for four days at our home on Long Island, N. Y. We depend on electricity for everything from well water to cooking and heating. A good snowstorm will also knock out the power.

The time has come to invest in a generator for these emergencies. Naturally, a whole bunch of questions arise. What size generator will provide for the very basic needs in a house? What is involved in connecting the generator to the house? Is noise a consideration? I know most of the units run on gas. Are any available that can tap into the fuel tank that runs the oil burner?

—Vincent Cafiero, Irvington, N. Y.

Ezra Auerbach, owner of Energy Alternatives on Lasqueti Island in B. C., Canada, replies: Storms like Bob quickly remind us just how dependent we are on grid-supplied power needs. When a storm knocks out power, it doesn't differentiate the essential loads from those that are merely conveniences; you are simply left in the dark.

There are two basic ways to back up grid-supplied electricity; one is with a generator, the other is with a battery-based electrical system (for more on battery-based systems, see *FHB* #62, pp. 68-71). In either case, it is probable that the backup system will power only selected circuits rather than the entire household electrical panel. While it is possible to buy a generator large enough to power an entire home, it's rather expensive. A more economical approach is to select the circuits that are essential and then size the generator for these circuits.

The biggest electrical hogs in homes are devices that heat, such as water heaters, space heaters and electric stoves. You will probably have to live without these circuits during a power outage. Essential loads are refrigeration, lighting, communications and furnace motors. None of these loads is prohibitively large, and all can be powered by a generator. A microwave is good for cooking during a power outage because of the small amount of current it requires. A woodstove is the best backup for heating.

Once the essential circuits are identified, they should be removed from the main distribution panel and wired into a subpanel. The subpanel is then wired to receive power either from the main panel or from a backup source. A manual transfer switch should be used as the method of selecting either grid or backup power. Don't be tempted to install an automatic transfer switch because of its apparent convenience. If an automatic switch fails, it can send power back down the grid, thus endangering line-repair personnel. Generally speaking, the subpanel should not be larger than 30 amps. A 30-amp panel can be powered by any generator over 4,000w. All of this work should be done by a licensed electrician in accordance with local electrical codes.

As for noise, almost all small gas generators run at 3,600 rpm, so they make similar amounts of noise. Some mufflers, however, are better than others. In general, the slower the rpm, the longer a generator will last and the quieter it will be. If noise is a concern, you might want to investigate the generators designed for motor homes. Although RV models are more expensive, they are usually quieter.

Many generators offer low oil pressure shut-off devices, as well as electronic ignition systems, features that make for simple and reliable generator operation. You are correct that most affordable emergency generators run on gasoline. I don't know of any designed to run on fuel oil, although there are generators that run on diesel fuel and on propane.

INSULATING A TIGHT SPOT

I have a lovely tongue-and-groove ceiling beneath a mansard roof that I need to insulate before converting the attic into living space. The space is, at best, only 2 ft. high, but I am loath to tear out the ceiling boards. What types of insulating products can you recommend for this job that will save the ceiling?

—Robert Eden, Hillsdale, Mich.

John Ross, an engineer with Westwind Corp. in Vienna, Va., replies: Fiberglass batts are almost as cheap as blown fiberglass, but they stay put and cover the area evenly. Also, kraft- and foil-faced batts provide their own vapor barriers. Although the space is too low for you to stand in, or even sit in, you can install fiberglass batts without going into the attic or destroying your cherished ceiling.

First remove about a 1-ft. width of the ceiling boards from the entire length of one edge of the ceiling (drawing on facing page). If you can, stick your head up there to locate the ceiling trusses; if you can't get your head up there, use a mirror. This is also an excellent time to check the condition of the roof. Then measure the height of the insulation space at the edge of the roof. The batts you install depend on the height available. However, even if you've only got 3½ in. of clearance at the edge, I suggest installing 6½-in. R-19 and compressing both ends of the batt. The difference between R-13 and R-19 over the entire ceiling will be noticeable in your climate.

On the end of the ceiling opposite from the removed boards, drill ¾-in. holes in the middle of each truss cavity. Run a snake through this hole to the side of the room where you removed the boards. Tie a length of clothesline to the snake and pull it to the open part of the ceiling.

After you've cut the insulation batts the same length as the truss cavities, trim a paint stirrer stick about 2 in. shorter than the width of the batt (12½ in. for trusses on 16-in. centers) and staple the stick across the front of the batt. Fold the paper and staple it to both sides of the stick. The stick has to be shorter than the full width of

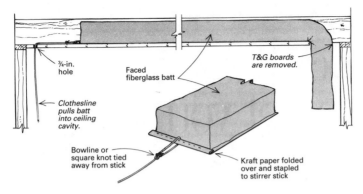

³⁄₄-in. hole

Faced fiberglass batt

T&G boards are removed.

Clothesline pulls batt into ceiling cavity.

Bowline or square knot tied away from stick

Kraft paper folded over and stapled to stirrer stick

the batt so that it won't get caught on the truss webs. If the batt has folded edges, leave them folded.

Make a small hole in the middle of the batt's paper facing, just behind the stirrer stick, pass the clothesline through the hole and tie a loop in the line about 6 in. in front of the stirrer stick. This loop has to be large enough so that you can tie the loose, or bitter, end of the line to it, yet the knot and the loop together must be small enough to pass through the hole in the ceiling. A bowline knot or a square knot works well because each is easy to untie.

With the slack taken out of the line, one worker feeds the insulation batt into the open end of the attic while the other worker gently pulls the line. The batt is fed so that the vapor barrier is face down in the attic. When the knot appears through the hole, the batt is in place. Untie the knot and pull out the line. Repeat the process for each truss space.

CONTROLLING SPA-ROOM MOISTURE

I'm getting ready to build a spa room like the one featured in the article "Adding a Craftsman Spa Room" *(FHB #58, pp. 58-61)*. My biggest concern is moisture control. Here in north Florida it gets very hot and humid. Will I need special ventilation? And with all the hot, moist air in a spa room, how do I keep the windows clear? I'm really concerned about the moisture problem down the road. What will the condition of the room be five years from now? I don't want the home owner complaining and having to spend more money later.

—*Kenneth Bowman, Lanark Village, Fla.*

M. Scott Watkins, a designer and builder in Arlington, Va., replies:
Like all major issues considered in the design and planning phase of a project, moisture control in a spa room must be integrated with

every other aspect of the project. Without an intimate knowledge of these other aspects, I hesitate to recommend any specific strategy for your project. I can, however, share with you some basic spa-room phenomena I have observed.

Excessive moisture in a spa room comes from three sources: water that splashes onto the floor when bathers exit and from the jet action, water vapor that saturates the air and condenses on interior surfaces, and water that leaks from the equipment when draining the spa or servicing the filter. None of these sources is continuous. When not in use, the spa should be covered or drained (primarily for heat conservation but also to prevent evaporation), and under most circumstances, servicing should occur only once every three months.

In developing your strategy for moisture control, ask the home owners how the spa will be used. Most people are satisfied with a 10-minute or a 20-minute soak (with spa water temperatures over 100°F, it is hazardous to remain in the tub any longer than that). In this case toweling the floor and opening windows after using the spa are adequate to vent the moisture on most days in your climate. If the owners use their spa this way on a daily basis, covering the spa with an insulating blanket (bubble sheet) or a custom-fitted insulated cover (vinyl over foam) should be part of their ritual.

If, on the other hand, the owners use the spa infrequently, say for entertaining once a month, they should drain the spa after each use. I would recommend consulting a local spa retailer/installer for further advice about spas and spa equipment and perhaps a building designer or architect for specific advice on detailing the construction to suit the Florida climate.

NAME THAT FUME

One of my best residential clients has a problem concerning the air quality in his home that I completely renovated in 1988. The house was gutted and refitted with insulated windows and doors, extra kraft-faced batts were installed everywhere possible, and Benjamin Moore primer/sealer and finish coats cover the walls and the ceilings. We made a conscious effort to avoid materials that might trouble the owner and his many allergies. The problem is on the first floor where we vaulted the ceilings and installed double- and triple-ganged skylights. On warm spring days, with no HVAC equipment active, a strange smell, something like new rubber floor mats or plastic trash-can liners, is very evident in these rooms. With the central air conditioning in use, the smell eventually

disappears. The odor is not a stale one that you would normally associate with stagnant air; the owner describes it as "an ozone smell."

I have been there on a bright, spring day and checked an area on the oak floor to see if the direct sunlight was affecting the stain and Glitza finish, but that wasn't it. The operable skylights were checked for mildew, and we closely inspected the attic and the crawl space for animals, vegetables, minerals, etc. It is odd to me that this problem only recently came up, whereas the owners have been living there for three years with no problem. What could be causing this odor?

—*Robert Jackson, Atlanta, Ga.*

Dave Menicucci, research engineer at Sandia National Laboratories in Albuquerque, N. M., replies: Most odors in residential buildings come from cooking, combustion, outgassing (the release of gases from solids or liquids), malfunctioning appliances or building systems and decaying organics. Cooking and combustion (especially from smoking) are the primary sources of pollution and odor in residential buildings. Cooking can produce strong odors that may last for days. Particulates from combustion may permeate fabrics and furniture, causing odors that can only be removed by washing.

Odors from decaying organic material usually originate in garbage cans; however, occasionally small animals may become trapped in an obscure area of a home and die. The animal carcass may be difficult to locate, and odors may linger for months.

Sometimes odors come from malfunctioning appliances or building systems, as well as from outgassing. Condensate drain pans under refrigerators that hold water for long periods of time cause mildew and mold odors. Sewer gases can creep into homes through traps that have dried out from nonuse. Pouring about a quart of water down the drain to refill the trap solves this problem. Outgassing occurs in new fabrics and wood products. Paints and varnishes may outgas for weeks after they are applied. Enclosed surfaces, such as the interior walls of cabinets, may outgas even longer. Resins used in fiberboard products also outgas for a long time. These outgases may also react with other gases to create different odors in the home. For example, the gases released from curing polyurethane may react with the odorant in natural gas to create an undesirable odor. Because the kitchen is the most likely room to have open gas flames, the odor may seem to originate there.

You can deal with an odor by eliminating its source after some investigative sniffing or by diluting the odorant with outside air from open windows and doors. These are two simple and cheap methods. Air filters, which remove particulates, and activated charcoal filters, which absorb organic gases and vapors, will eliminate airborne agents that cause odors, but these systems are often expensive to install and maintain. Finally, masking the odor with a deodorant is the least desirable solution because it adds a new gas to the environment, which may cause problems.

Your client says the problem has only recently occurred, even though the construction was completed three years ago. This suggests that the odor is not related to the building materials you used. Outgassing from these materials would have been most intense just after construction and would have been noticeable immediately.

Second, you say the odor smells like rubber or plastic trash-can liners. Both of these odors are typical of outgassing plastics or treated fabrics.

Third, the odor is apparently more acute on warm days. Heat can accelerate outgassing from new furniture or from decaying organics, both of which could have been added to the environment recently.

Fourth, you say the odor disappears soon after the ventilation system kicks in. Occasionally, a very dirty electronic air cleaner will produce electric arcing that creates ozone, an odoriferous gas. Also, insects sometimes crawl onto the high-voltage plates of the air cleaner and are burned, creating short-lived odors throughout the house. Obviously, the existing air-conditioning system and the electronic air cleaner are not sources of the problem.

Based on the evidence, outgassing seems the most likely source of the odor. I'd ask your client if he's added something new to the environment, such as furniture or treated fabrics. If so, he should temporarily remove the new material to see if it is causing the odor.

If this doesn't solve the problem, then additional investigation should be conducted. You may need to consult an environmental expert to test the air in the house.

ELECTRIC RADIANT SLABS AND HEALTH

I have heard that exposure to weak electrical fields, such as those from overhead transmission lines, home appliances and even electric blankets, poses health risks over time.

Do the wires embedded in an electric radiant slab give off an electrical field, or does the lead sheathing on the wires block such emissions?

—Alan Berry, Palos Verdes, Calif.

Richard D. Watson, research chairman of ASHRAE Radiant Heating and Cooling Technical Committee, replies: All energized electrical devices emit an electromagnetic field (EMF). Research, to date, has not found whether the relationship cited in studies of EMF and health problems is coincidental or causal. The lead sheathing on wires in an electric radiant slab is for grounding purposes and has no impact upon EMF.

If you're concerned about EMF, bear these two facts in mind: EMF intensity, measured in milligauss, decreases geometrically (the inverse square law) as you move away from the source of the EMF. And EMF intensity increases with the amount of power (wattage) in a circuit. Information concerning power output is available from manufacturers of electric radiant-heat products. Both private outfits and many local utilities will measure EMF levels in your home. However, no standardized threshold of safe vs. unsafe levels of EMF has been established.

STANDING UP TO SALT WATER

I've just become property manager of a house located on the windward side of an island in the Bahamas. The house is only 30 ft. from the shore, so its exposure to salt water is tremendous. Paint lasts only two years, flashing and hardware disintegrate much faster than they would inland, and wood is quickly eaten by termites. Although I'm an experienced builder, the salt-water environment is new to me. Any tips on maintenance?

—C. Bradley Bush, Boulder, Colo.

Bermuda contractor Scott King replies: Unfortunately, there are no simple answers to your maintenance problems. Living on an island is like living on a boat—high-grade materials and careful preparation are crucial. Materials that hold up well in wet, salty environments include acrylic latex primers and paints; aluminum or PVC windows and doors; and brass or stainless-steel hardware. Meticulous preparation of surfaces to be painted, whether masonry or wood, will give paint jobs a longer life span. Here in Bermuda, we might get four to six years out of a first-class paint job—longer if we perform routine maintenance on a yearly basis. Even PVC or aluminum windows and doors require periodic lubrication and care to keep their mechanisms from corroding into useless junk. To build exposed wood surfaces we use pitch pine, a particularly hard, decay-resistant yellow pine. For hidden surfaces the pressure-treated stuff is even better. A good pest-control company can get rid of the termites; they may also offer a monthly checkup service.

The common thread in all of this is that in a salt-water environment, regularly scheduled maintenance and quality materials can save you a lot money in the long run.

DESIGNING SOLAR STORAGE

I want to build an active-solar heating system with long-term thermal storage. I know of one individual who built such a system using two 6,000-gal. stainless-steel wine casks that were insulated and buried outside his house. The sun heats the water during the warmer months, the water releases heat throughout the winter via a radiant system.

As an alternative to buying such tanks, I'm considering using my basement as a large cistern. I would insulate it and line it with a pool liner. I would seal the floor joists above this area with 6-mil or heavier plastic, along with a combination of rigid and fiberglass insulation.

Do you think such a system is feasible? Would the moisture present a problem if the steps were taken to seal off the basement area?

—Michael S. Niziol, Apalachin, N. Y.

Marc Rosenbaum, P. E., of Energysmiths, Meriden, N. H., replies:
Building a basement-level tank to store solar-heated water is both practical and feasible. By using one end of the basement, you only need to pour one extra wall, since three of your "tank" walls are already there. Pouring the fourth wall 18 in. to 24 in. shorter in height will leave access for construction and maintenance. You should consult a structural engineer before pouring the new wall: you want to make sure that it's strong enough to withstand the lateral loads imposed by the water. Stress on the slab shouldn't be a problem, though. As long as it sits on compacted soil, it should easily support the 375 lb. per sq. ft. weight of 6 ft. of water.

In order to prevent thermal bridges to the ground, you'll need to insulate the inside of the tank with 4 in. to 8 in. of extruded polystyrene foam. The exact thickness will depend on the various sizes of the collectors, the tank and the house. Larger tanks have a lower surface to volume ratio and therefore will lose heat more slowly.

A standard pool liner is probably inadequate because of potentially high water temperatures. Instead, you may have to make your own liner from a synthetic rubber membrane. Make sure that the membrane will hold up over time to temperatures of 180°F to 200°F. You'll have to call several manufacturers of synthetic rubber membranes to see if their product is suitable for such an application.

Install the liner membrane over the foam. Lap it over the top of the foam and attach it to the concrete walls. For moisture control, stretch a separate membrane over the top of the tank. Seal this to the liner membrane in such a way that it can be removed if the need arises. You can do this by lapping the two membranes, then using 2x4s and lag screws to fasten them to each other and to the wall. The exact method isn't important, however; the point is that you may need to remove the cover someday, so you don't want to glue the top membrane to the liner membrane. At the same time, you want to keep potentially harmful water vapor inside the tank. Insulate the top membrane by covering it with unfaced fiberglass batts or rigid foam. Foam is easier to work with, while fiberglass is cheaper.

To avoid potential leaks, make sure that all pipe penetrations are at the top of the tank; there should be no penetrations below the water line. A drain placed in the concrete slab beneath the tank will serve as an escape path for water that has leaked from the liner. This is not the way to drain the tank, however—you'll need to do that with a pump.

Where you live in the Northeast, winter sunshine can be sporadic. I recommend that you install an annual storage-cycle system, where most of the energy required for winter heating is collected during the summer. This will require a large tank (on the order of the wine-cask system you mentioned, if not larger), as well as careful optimization of house insulation levels, collector area, tank size and distribution system design. To ensure the success of such a complex system, you should consult a qualified solar engineer.

INSULATING BENEATH A SLAB

Standard preparation underneath a residential slab these days seems to include the use of rigid insulation. But I've heard that with the perimeter wall insulated to below the frost line, the temperature of the earth below the slab is about 50°F, so how important is it to insulate against such mild temperatures?

—Brent Harold, Wellfleet, Mass.

William A. Randall, an architect in Eugene, Ore., replies: You are right that the earth below the frost line does maintain a relatively constant 50°F. The primary goal of insulating floor slabs is to prevent the escape of heat at the perimeter, not at the center of the slab. Generally, heat loss through the center of the slab will merely warm the earth slightly and thereby reduce heat loss until equilibrium is reached. The net effect is virtually no heat loss through the center of the slab.

Most codes allow the option of insulating the perimeter of the slab to below frost line (i.e., insulating the foundation wall to the bottom of the footing) or returning insulation under the slab a specified distance (usually 24 in.). I generally choose the former to allow a sound base under the slab for bearing. This effectively prevents heat loss from the slab perimeter and maintains energy-conscious design.

DESIGNING A ROOT CELLAR

I want to build a root cellar or cold room in my basement. The space available includes a cement floor and one north-facing exterior block wall that extends about 18 in. above grade. Where do I place the insulation and vapor barrier? What are the ventilation requirements?
—Charles Milburn, Islington, Ont.

Marc Rosenbaum, P. E., of Energysmiths, Meriden, N. H., replies: Building a basement root cellar is both practical and feasible in a cold climate. Your goal should be a room that stays within the 32°F to 40°F range, with humidity between 70% and 90% (I haven't found low humidity to be a problem, but if it is, raise the humidity by filling a couple of 5-gal. pails with sand and watering them periodically).

In new construction, I try to design root cellars into a corner of the basement on the north side of the house, with a minimal amount of the concrete wall exposed above grade. The concrete should be heavily insulated on the exterior with extruded polystyrene foam (in Canada, sheets of rigid fiberglass insulation are available; use these instead for lower environmental impact). The foam should extend 4 ft. below grade (drawing on facing page). No insulation is needed on the bottom half of the wall. Although an old saw about root cellars says that you need a dirt floor to maintain high humidity, my experience indicates otherwise (besides, dirt could serve as a radon pathway), so I continue the slab into the root cellar. As with the rest of the basement, the slab in the root cellar is poured over 4 in. to 8 in. of crushed stone and a polyethylene vapor retarder. The only difference is that I omit any under-slab insulation in the root cellar.

Before pouring any concrete, I build a box around the perimeter of the root cellar with pressure-treated 2x lumber placed on edge. I then pour the slab so that its surface is flush with the top of the 2x. Not only does the 2x thermally isolate the root cellar slab, but it also serves as a nailer for the plates of the two framed walls.

The framed walls and ceilings of a root cellar should be well insulated. I build 2x6 walls and insulate them with fiberglass batts; the house floor above is usually 2x10 or 2x12 framing, which I insulate with fiberglass batts as well. I don't put in a vapor retarder, but I do use ½-in. plywood on the inside of the root cellar to make

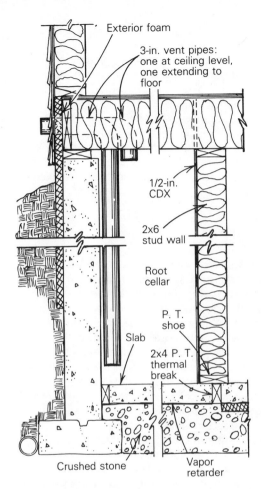

Exterior foam

3-in. vent pipes: one at ceiling level, one extending to floor

1/2-in. CDX

2x6 stud wall

Root cellar

P. T. shoe

Slab

2x4 P. T. thermal break

Crushed stone

Vapor retarder

sure that the walls are airtight. For access, I prefer a well-weatherstripped, insulated steel door.

To keep the space cool, I passively ventilate it to the outdoors with two 3-in. PVC pipes. One pipe ends near ceiling level in the root cellar; the other drops down to the floor. The varying heights create a thermosyphon loop, letting the warmer air exit the root cellar at the top while colder air from the outside falls to floor level. In my own root cellar, I plug these pipes if the temperature drops below -10°F to prevent freezing. A more automatic way would be to ventilate with a thermostatically controlled fan, but I haven't tried this.

If you don't have two concrete walls to use (for example, if you have to frame the root cellar in the middle of one wall), you may want to dampen temperature extremes by adding extra thermal mass. If the room isn't full of produce, you could increase thermal storage with plastic jugs filled with water.

All stored food should be raised off the floor. For instance, one wall of our root cellar supports 20-in. deep, vinyl-coated, steel-wire shelving from floor to ceiling. We find that beets and carrots keep well if we double-bag them in large, perforated poly bags. Onions and potatoes we just store in boxes. You'll find that various types of produce differ in their requirements for temperature, humidity and air. Two good sources of information are: *Root Cellaring* by Mike and Nancy Bubel (Rodale Press, 1979) and *Putting Food By* by Ruth Hertzberg, Beatrice Vaughan and Janet Greene (Stephen Greene Press, 1988).

DRYING A WET SLAB

The house I bought last year has a moisture problem in the concrete slab. Although there is no actual water or evidence of there ever having been any in the 10-year life of the house, the entire downstairs feels damp. Any impervious material that's left overnight on the bare concrete will have a dark patch under it the next day.

About two-thirds of my perimeter foundation is set into an embankment. But there are no water problems around the house, and the two sections of foundation drain I uncovered as a check were both below the level of the slab and completely dry. Common sense, then, suggests to me that I have no vapor barrier under the slab.

My problem is the contradictory information I've received on what to do. One waterproofing contractor told me that I have hydrostatic water pressure under the slab, that the pressure will eventually force water to enter, that no protective coating exists to stop it, and that nothing short of expensive foundation drains installed under the slab will solve my problem. Another one told me that all I have is a vapor problem, that foundation drains won't help, and that coating the foundation with a penetrating sealer will keep it perfectly dry. Needless to say, I'm more confused with the contradictory solutions than with the original problem. Any advice?

—Charles Watkins, Bainbridge Island, Wash.

David Benaroya Helfant, a structural and enginnering inspector from Berkeley, Calif., replies: The problem that you describe is a common one in subgrade habitable rooms and in crawl spaces that are excavated into slopes. I agree that your slab was probably poured without a vapor barrier, and that it probably wasn't engineered to deal with the site's obvious water problems.

Moisture build-up in the slab may be attributable to one or more of the following: the perimeter drain, even though it's below the level of the slab, is obviously not picking up everything. This may be a sign that water is percolating up from below. Water may also be backing up from a lower level on the slope. Or surface runoff may be seeping below the level of the drains and emerging in the area of the slab (an artesian well is a good analogy, because it's a classic example of water that emerges at a strange elevation). While it sounds as if leakage of this sort is the main culprit, condensation due to the different temperatures of the slab and the air may also be a cause. Finally, the lack of proper air ventilation and circulation can contribute to less than desirable evaporation rates.

The potential remedies available are varied. The most thorough—and most expensive—approach involves busting out the entire slab, over-excavating by 6 in. to 8 in. and laying a 4-in. to 6-in. blanket of free-draining rock (drawing below). Perforated collector pipes are embedded in the rock and tied into the perimeter drain system, which should drain well below the building into a suitable city storm-water depository. A thick, impermeable vapor barrier is laid over the drain rock, followed by 2 in. of clean sand. A steel-reinforced, 4-in. concrete slab is poured on top of this.

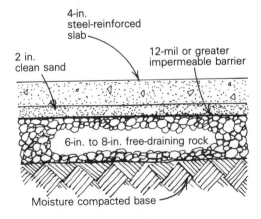

4-in. steel-reinforced slab

2 in. clean sand

12-mil or greater impermeable barrier

6-in. to 8-in. free-draining rock

Moisture compacted base

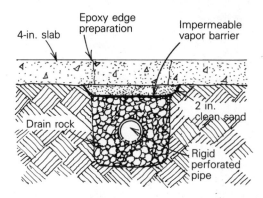

Epoxy edge preparation

Impermeable vapor barrier

4-in. slab

Drain rock

2 in. clean sand

Rigid perforated pipe

A second approach is to let in drains in the slab by saw-cutting the slab in strategic locations and installing a field of perforated pipe that's connected to an outlet point below the home (drawing above). Like the moisture-barrier slab, the drain field would have an impermeable, puncture-proof vapor barrier protected by two inches of clean sand. This would be covered with concrete that's finished to match the existing slab.

A third approach is to install a dry well or sump pump in a strategic location to de-water the area beneath the slab. A fourth approach would be to install a rigid channelized polyethylene matrix on top of the existing slab (after waterproofing the existing slab) and to pour a new slab on top of this. Emerging water from the old slab would flow through the channelized material to a collector, instead of wicking up through the new slab.

Finally, you might coat the walls and floor with a traffic-duty rated waterproofing sealant. The likelihood of this being successful is about 50/50, depending on the engineering specs of the material. The coating would have to be renewed periodically to maintain its seal.

I have listed these solutions beginning with the most expensive and ending with the least. The probability of long-term success is directly related to expense: the first solution has the best chance to succeed; the last has the least. In order to keep repair costs to a minimum, however, you should probably try the last option first.

READERS REPLY

I'm forced to take exception to David Helfant's advice. The drastic measures he recommends are akin to prescribing brain surgery as the first line of attack for a headache, when a few aspirins might do the trick. As a former home inspector, I've seen many wet basements. Over 90% of them are caused by too much surface water too close to the house. The solution is to make sure that water drains away from the house. To do so, you should keep gutters and downspouts clean; use long splash blocks or underground downspout extensions to provide an easy drainage path away from the foundation, and slope all the ground within 6 to 8 ft. of the house down at least ½ in. per foot of horizontal run.

If these measures don't work, you've spent very little money and can then consider the more extreme and expensive measures. Consultation with a competent home inspector can be a real help in most wet-basement situations.

—M. B. Williams, Potomac, Md.

MOISTURE PROBLEMS IN THE CRAWL SPACE

I own a single-story brick home, built in 1951. It has 4-in. cinder-block walls with a brick face and uses a post-and-beam support structure. The windows are of an old steel-casement type. We have been using window air-conditioners to cool the house in summer.

Before we bought the house, the crawl space had been insulated with fiberglass between the beams, and the ground was entirely covered with plastic. Earlier this year I had occasion to go into the crawl space and discovered that the insulation beneath the two air-conditioned bedrooms was soaking wet. Some of it was so heavy with water that it had dropped out from between the joists. The wood was wet, and there was mold growing on it.

One suggested remedy was to staple a layer of polyethylene plastic to the underside of the floor above any new insulation. But I'm afraid that this would cause water to condense in the wood sub-flooring. The only other solution I can think of is to pull up the carpets and cover the floor with an oil-base varnish, which would then act as a vapor barrier from the inside. Do you have a better solution?

—Henry J. Stock Jr. Bethesda, Md.

David Kaufman of Phoenix Energy Services in Waldeboro, Maine, replies: The moisture in your joists and insulation is probably coming from underneath your floor, not from above it, unless condensation from the air-conditioning units is leaking down to the crawl space. In either case, though, adding varnish or polyethylene to the floors won't solve your problem.

In the summer, the underside of the air-conditioned rooms is the coldest spot in the crawl space. Moisture is somehow entering the area and condensing there, much as water condenses on the outside of a cold can of beer on a hot summer day. The plastic ground cover makes it unlikely that the ground is the source of moisture, so I suspect the culprit is humid summer air entering through the crawl-space vents. This once-standard practice has lately come under increasing fire. Crawl-space vents were helpful in drying the wood in cellars of older houses plagued by annual interior spring floods. But it seems that introducing warm, humid summer air into an otherwise dry crawl space can actually increase moisture levels.

I recommend temporarily removing the insulation from the areas with moisture problems (to facilitate drying) and ventilating through the fall or taking other steps to get the wood dry. Then you can reinsulate the joist bays and close off the crawl space—no more ventilation. Some authorities even recommend winter ventilation to reduce moisture levels, but because of its impact on heating bills, I would suggest taking this measure only as a last resort.

Before closing off the crawl space, though, you should test for radon. If you do have a radon problem, then you'll have to deal with that separately. You won't be able to eliminate the crawl-space ventilation because that could result in a dangerous build-up of the gas. In this case, you'll have to try extending the polyethylene ground cover up onto the walls and installing an air barrier (like Dupont's Tyvek or Remay's Typar) across the bottom of the joists to curtail air flow into the insulation. Although polyethylene would work a bit better for this, it would also act as a wrong-side vapor barrier in the winter and once again trap moisture in the insulation.

RADON IN THE CRAWL SPACE

My wife and I are purchasing a house with a full poured-concrete basement. The house has a family-room addition, built over a dirt-floor crawl space that vents into the basement. The basement recently tested at 4.01 for radon. I plan on painting the main basement floor to seal out the radon, but I am sure the majority of the radon seepage is coming from the crawl space. I'm not sure the crawl space could be isolated, insulated and vented to the outside without becoming a major project. It seems to me a heat-recovery ventilator might be the easiest solution to the radon problem.

Neither the main house nor the family-room floor joists have any insulation or vapor-barrier covering. If I insulate the basement joists with fiberglass batts, would that reduce the radon level in the house? Would the fiberglass tend to collect and retain radon? Also, the house (built in 1976) has an oil-fired forced-air heating system with return cold-air ducts. Need I worry about basement air leaking into the forced-air system?

—Steve Malavarca, West Orange, N. J.

Charles Lane, a civil and environmental engineer in Shoreview, Minn. replies: You indicated that your house recently was tested for radon and the results were 4.01. Although you didn't say, I will assume that the measurement units were in picocuries per liter (pCi/l), which are the units used in the home radon-testing kits. If the units of measure were Working Levels, a reading of 4.01 would be dangerously high and the house should be evacuated. Also, the length of time during which the radon measurement was obtained will affect the results. Generally, a long measurement period (three months to one year) is best.

If my assumption of units is correct, the radon "problem" in your home is relatively small. The U. S. Environmental Protection Agency recommends a level of 4 pCi/l as the practical (not necessarily "safe") limit to achieve in homes. Therefore, before you invest in any major radon-mitigation strategies, I recommend that you install an alpha-track radon detector (one manufacturer is Terradex Division of Tech/Ops Landauer Inc., 2 Science Rd., Glenwood, Ill. 60425; 312-755-7911) in your basement and measure for a period of one year. They cost around $25, which includes the lab analysis. My guess is that the results will be below 4 pCi/l.

If, however, you are uncomfortable with the radon level in your home, you can seal the crawl space from the soil with a sheet of EPDM membrane (the material often used on commercial roofs). The edges of the membrane should extend up the walls in the crawl space and be sealed to the walls by caulk and a continuous wood strip nailed to the wall. Make sure that any seams in the membrane are properly sealed.

Sealing the basement floor with paint is one of the least effective techniques for radon mitigation. Installing insulation and polyethylene will not help either; they should be used to control energy costs and migration of moisture vapor, not to control radon. You shouldn't be concerned about radon attaching to the fiberglass or entering your forced-air system. Under some conditions, heat-recovery ventilators are used to mitigate radon. But in your case the level of radon is so low that adding a heat-recovery ventilator would probably not have a significant enough effect to justify the expense.

Additional information on radon reduction can be found in a pamphlet entitled "Radon Reduction Methods, A Homeowner's Guide," available for $1 from the Superintendent of Documents, U. S. Government Printing Office, Washington, D. C. 20402. You can also contact your state radiation office or nearest EPA regional office.

DRYER AIR INTAKE

I am building an energy-efficient house that has an air-exchange system, a direct-vent oil-fired boiler and a zero-clearance fireplace that draws outside air. With kitchen and bathroom venting part of the air-exchange system, the only device in the house that can upset the balance between intake and exhaust is the clothes dryer, which takes air from inside the house and sends it outside. I suspect that the negative pressure created by the dryer may inhibit the air exchanger's operation. Are you aware of any manufacturer of direct-vent dryers? If not, do you have any suggestions that could help me solve this problem?

—David D. Huntoon, Ridgefield, Conn.

Marc Rosenbaum, P. E. of Energysmiths, Meriden, N. H., replies: I assume, because you are heating with oil, that the dryer is electric, not gas. You are correct in recognizing that a clothes dryer acts as an exhaust fan, typically drawing 100 cu. ft. to 125 cu. ft. of air per minute out of the house. In a tight house this will create lower air pressure inside the house than outside. This difference in pressure may overpower the chimney draft of any naturally vented combustion device (such as a furnace, a boiler, a water heater or a wood-burning appliance), pulling flue gases back into the house. This action is called backdrafting, and it's potentially lethal.

It appears that the two combustion appliances in your home have sealed combustion, meaning that they get air for combustion directly from outdoors. The fireplace will need glass doors to seal it from the home. Sealed combustion devices have their combustion process completely decoupled from the air in the home, so they are not affected by pressure variations in the home and cannot be backdrafted.

If no danger of backdrafting exists in the house, I wouldn't be too concerned about the dryer causing problems for your heat-recovery ventilation system. It might throw the system out of balance, but only for the few hours a week that the dryer is in use.

I am not aware of any dryers on the market that draw air directly from outdoors. However, you could locate your dryer in a small room that has outdoor air supplied to it and is sealed off from the rest of the home. In this case you need an air inlet with a good backdraft damper so that the air inlet won't leak cold air when the dryer isn't in use.

Another option might be to have a sheet-metal shop make up a fitting that covers the dryer's air inlet louvers (my dryer has them on the back). This sheet-metal fitting could have a pipe stub on it that connects to your outdoor air ductwork. Keep in mind that any duct carrying cold air should be insulated. The duct should also have a well-sealed exterior vapor barrier to prevent condensation.

Using cold outdoor air in the dryer may degrade its drying performance. A solution would be to fabricate a concentric pipe-within-a-pipe to serve as both air inlet and exhaust. The inlet air would be warmed by the exhaust air, like a miniature air-to-air heat exchanger.

DRYER VENT THROUGH ATTIC

Our washer and dryer are on the second floor of the house. The dryer was originally connected to the same duct as both bathroom vent fans; flexible dryer hose vented to the soffit. However, condensation collected in the low spots of the flexible hose. Recently, we have simply vented the dryer into the attic. This is unsatisfactory because of the lint and because the hot dryer air condenses when it hits the cold attic air. It literally rains on the insulation.

What's the best way to run a dryer vent through the attic, and how do we cap the vent to prevent external elements from entering the house?

—Judy Sutton, South Windsor, Conn.

Mark Rosenbaum, P. E. of Energysmiths, Meriden, N. H., responds;
First of all, wherever possible, do not vent the exhaust from either a dryer or a bathroom into or through the attic or any other unheated space in the home. These exhaust streams are full of moisture and hence have a high potential for condensation. Also, the dryer and the bathroom fans should be vented separately and should never be vented out the soffit vent. Soffit vents are intake vents, so any air exhausted at that point will likely be sucked back up into the attic.

I assume that your laundry is in an interior space and venting directly through an exterior wall is impossible. Vent the dryer with 4-in. dia. metal duct, going vertically from the dryer through the second-floor ceiling. Seal the ceiling penetration so that the duct can expand in length as it heats up during use (a standard roof boot, such as you'd use to flash a vent pipe, would work). Continue the ductwork horizontally through the attic to the closest exterior wall.

Pitch the attic ductwork back toward the dryer so that any condensation will run back toward the heated space. Support the ductwork adequately to maintain this pitch—don't allow the ductwork to sag in the middle. If you come off of the dryer with a tee connector instead of an elbow, condensate will run back to the lower portion of the tee instead of back into the dryer. This will give you a place to inspect whether condensate was indeed forming and to drain the water away if necessary.

Penetrate the wall and use an Energy Saving Dryer Vent (Heartland Products, P. O. Box 777, South Kathryn Road, Valley City, N. D. 58072; 800-437-4780), which is designed to provide a tight seal when the dryer is not in use. Follow the dryer manufacturer's recommendation for maximum allowable duct length. My Sears dryer, for instance, allows a maximum of 44 ft. of 4-in. duct with two elbows, and 35 ft. of 4-in. duct with three elbows.

The ductwork should be taped with a high-quality duct tape. Tape both longitudinal and transverse seams. Do not use sheet-metal screws to connect the ductwork because the screw tips protrude into the duct and will collect lint. Insulating the duct with at least 1 in. of fiberglass insulation will keep the duct warm during dryer operation and will minimize condensation on the duct walls. The more insulation you install, the less likely you are to have condensation in the ductwork. Insulation that's carefully wrapped around the duct will work better than fiberglass batts just piled on top.

Because the dryer vent's air seal to the outdoors cannot be made between the heated space and the unheated space (the second-floor ceiling), natural convection in the ductwork when the dryer is not running may still convey enough water vapor to the attic ductwork to cause condensation. Keeping wintertime relative humidity levels in the house to 40% or below will minimize this problem.

HYDRONIC COILS FOR FORCED-AIR HEAT

I am renovating a house that was built in the 1920s. The new floor plan requires lots of changes to the original layout. As a result, I have had to remove all the hot-water baseboard heating on the second floor. I will be adding ductwork in the attic for heating and cooling. I would like to keep the high-efficiency boiler and add a water-to-air heat exchanger to satisfy the heating requirement of the house.

Can you provide me with some details as to where I might find a water-to-air heat exchanger? Also, to prevent freezing I intend to add antifreeze to the system. Is there a separator (check valve) that can be added to the boiler system to prevent antifreeze from backing up into the house water? Can you comment on the pros and cons of this heating concept? Is there a better solution with forced-air heating?

—Joseph Philip Green, Friendship, Md.

Denny Adelman, a hydronic consultant and designer in Newport, R. I., replies: In-duct hydronic coils for forced-air heating are a standard item in supply houses. They resemble car radiators and can be installed easily in existing ductwork. They need to be sized to the heating load and the fan/duct capacity.

Air-handler units, which are also commonly available, incorporate fan, hydronic coil and air-conditioning coil in one factory-prepared box that can be inserted into ductwork with minimal installation costs.

Most localities require installation of one (or even two) backflow preventers at the fill point of a heating system. If yours does not have one, this is easy to fix. Also, please note that the recommended antifreeze for hydronic systems uses nontoxic propylene glycol as its active agent. This is the same propylene glycol that is used as artificial flavoring in supermarket ice cream. Never use automobile antifreeze (ethylene glycol) in a hydronic system.

Using an efficient boiler to power a hydronic coil is the best way to design a forced-air system. Because the primary heat source to your house is 190°F water (and air) instead of 500°F air, the fan cycles are longer (more comfortable) and the air is not as dry (healthier).

When compared with hydronic heating, however, forced-air heating in whatever form is still relatively uncomfortable, inefficient and noisy. There are some very handsome European panel radiators for heating water that do not need to ring the house perimeter to work. You might find that you can keep both your boiler and your superior hot-water heat.

EXHAUST VENTS IN A TIGHT HOUSE

I have a couple of questions about a Cape Cod-style house I am planning to build in upstate New York. For a windowless half-bath, can an exhaust fan be vented up into the attic space or down into the basement, or must it go through the roof to open air? Also, I dislike punching a hole through my projected R-19 wall to vent a clothes dryer. In the home I live in now, I disconnect the exhaust hose from the vent pipe in the winter, attach a lint-catcher to the end, and let the warm moisture exhaust into the uninsulated basement, with no problems.

—Burton L. Hotaling, Wooster, Ohio

Energy consultant John Hughes replies: Bathroom fans should never be vented into attic space, and because they're intended to carry off odor as well as moisture, I wouldn't vent one into a basement, either. Instead, dump the moist air outside.

Much the same can be said about venting clothes dryers. The amount of moisture that they carry off is enough to cause severe moisture and condensation problems in a poorly vented attic space. While venting one into an old, leaky house might not cause moisture problems, it very well could in a new, tight house. Dump it outside as well. Have your contractor carefully cut a hole in the house wall, run the vent pipe through the wall and use some spray foam to seal up around the hole. If the work is done right, the vent passage will be just as tight as a normal wall section.

COOLING EFFECTS OF RAMMED EARTH

I bought a lot in Fountain Hills, Ariz., and expect to build in about a year. I'm interested in the article on rammed-earth construction in *FHB* #34, as well as the "Earth and Sky," article by Brian Lockhart *(FHB* #39, pp. 49-51). What is the calculated effect of rammed-earth construction, compared with that of conventional construction, on cooling in the hot season? I suspect that a rammed-earth house would be easier to cool because of the mass of the walls.

—*Clair D. Siple, Lakefield, Minn.*

Architect and author Brian Lockhart replies: To answer your questions, I consulted Tom Schmidt, of Schmidt Builders in St. David, Ariz., and Tom Wuelpern of Rammed Earth Construction in Tucson. Both have extensive experience with earth-building in a range of different climates, and both use rammed-earth walls in superinsulated, passive-solar homes. While the cost is higher than for stick framing, it's cheaper than other mass-wall constructions. The temperature in these homes varies only slightly; the storage mass radiates or absorbs heat or cooling, depending on the time of the year, to give an even comfort level.

Conventional construction without this storage capacity needs a more constant input of heating or cooling. A home with rammed-earth walls can be comfortable both in summer and winter if designed for its environment. The Phoenix area, where Fountain Hills is located, has a far greater requirement for cooling than for heating—3,500 cooling degree days versus 1,550 heating degree days. To isolate the storage mass from the high outside temperatures, the house must be well insulated on the exterior. Windows must be of good quality, and at least double-pane. Window openings, unless shielded, should be minimized on the north, east and west, as well as overhead, to reduce summer solar gain. In summer, we receive minimal direct gain through south-facing windows. In winter, however, south-facing glass can be an invaluable solar-heat source.

With average day and night temperatures above 80°F from May to September (91°F in July), the home will need some mechanical cooling. The best solution is to use air-conditioning at night. The stored cooling can carry the house through most of the day, and because the unit is not working in the extreme daytime heat, it won't hurt the electric bill so much.

HOT-SURFACE CONDENSATION

A home owner has asked me to cure a condensation problem at the peak of the cathedral ceiling of a 16-year-old post-and-beam home built with stress-skin panels. The panels consist of 3½-in. urethane foam insulation between plywood skins. The roof is not vented. The problem is a peculiar one, in that the condensation appears only during humid summer weather. I always thought condensation formed on cool surfaces, but this is the hottest area of the house. When the air-conditioner and a fan (at the upper level, near the ceiling) are running, the condensation subsides or disappears.

There was no condensation problem until a few years ago, and as far as the owner knows, nothing has changed to cause it. I'm at a loss as to how to correct it. It does not seem to be caused by a leak because the condensation forms along the whole length of the house and in an ell, and the ridgecap has been replaced twice. The house is in Norwich, Conn.

—Robert F. Knight, Hopkinton, N. H.

C. K. Wolfert, vice-president of technical services at Air Vent, Inc., replies: I think you're dealing with a condensation problem, not a leak. Condensation will occur any time the air pocket at the peak is hot and moist enough to have a dew point that's higher than the temperature of the structural mass. If the air at the peak is 100°F with a relative humidity of 70% (feasible at the peak of an unvented cathedral ceiling during hot, humid weather), then its dew point is 90°F. In other words, in this instance, the moisture in the air will condense on any surface that is 90°F or less (which could describe the ceiling, given the panels' tendency to stay cool).

Lowering the temperature or relative humidity of the air will lower its dew point. When the air-conditioner is used along with a circulating fan, the air is cooled to a temperature of about 85°F and dried out to a relative humidity of 45%. This lowers the dew point to about 60°F, which would be lower than the temperature of the structural mass at the high areas. This can be checked with a thermometer and a hygrometer.

DAMP MASONRY

My log cottage, built in 1956, has an impressive fieldstone fireplace and chimney. We have added a woodstove in the fireplace, with a stovepipe that runs up the chimney. In the past few years I have noticed with growing frequency that the stone hearth becomes dark due to dampness. The darkness disappears during dry days and reappears four to six hours after a rain storm. No water appears on the interior vertical surfaces. I've checked all roof flashing and cannot find any leaks.

One particular wet weekend, after several days of rain, I arrived to find virtually all of the interior chimney discolored with dampness, and it was too cold for this to have been condensation. It seems that water is entering the entire chimney structure, possibly through cracks or around the stones between the mortar.

What might be causing this problem, and what's the most effective way to cure it?

—Neil Haist, Willowdale, Ont.

Stephen Kennedy, a stonemason in Kettle, Kentucky, replies: Not having seen the chimney and fireplace in question, I can only make an educated guess as to the origin of the wetness and the ways to prevent it. But I would check all the flashing again. In uncoursed or random stonework, it is more difficult to do a proper flashing job because there are no continuous houizontal joints to tie flashing into, as there would be in a brick, block or coursed-stone chimney.

Assuming that the flashing is okay, the next most obvious trouble spot would be where stonework meets logs on the sides of the chimney. Wood and masonry join poorly in the best of circumstances. Often mortar is used to seal against wood and initially appears to join well. However, it will separate from the wood as the wood dries and contracts. Additional flashing and caulking may be required here.

The juncture of your metal stovepipe and concrete chimney cap may be a problem spot, especially if the cap is not sloped outward, and water is allow to puddle against the stainless steel pipe. I wouldn't trust a mortar joint against steel to do a proper job of sealing, as the steel will expand and contract slightly. I'd seal the stovepipe with an approved high-temp caulk.

Also, make sure that the concrete cap does slope outward slightly. A good cap also extends beyond the chimney below and acts as a roof for the masonry, shedding a high percentage of the water before it can find other openings.

Another possible source of moisture infiltration is the porosity of the masonry itself. I would go over the whole chimney, filling with mortar any gaps that were more than ¼ in. wide and filling smaller gaps with clear silicone caulk. Then one coating of clear masonry sealer should do the rest of the job. Applying the sealer should be the last thing done, after the larger gaps are sealed.

What interests me about your question is the implication that the moisture has only recently appeared, in a 35-year-old chimney. Was something new done to it in the last few years? For example, was the roof replaced? How new is the steel chimney? An earthquake would raise hell with a fieldstone structure. Perhaps you had a small earthquake that set up some cracks. The cumulative effects of acid rain could be another culprit, slowly dissolving away lime from the mortar. If acid rain is the only problem, the openings in the masonry would be tiny, and a masonry sealer might be all you'd need.

HIGH-VELOCITY AC SYSTEMS

I plan to install a central air-conditioning system in my home. The house was constructed with hot-water baseboard heating, which means that I must have air ducts installed.

I've had a few local air-conditioning vendors inspect the house. The typical installation they offer consists of a compressor (outside the home) feeding a large air-handler/duct system in the attic, from which 10-in. to 12-in. dia. pipes drop through second-floor closets into the first-floor ceiling.

I've heard of a system that uses 3-in. to 4-in. dia. flexible hose in place of the larger ducts. Evidently the 3-in. system operates on a high-velocity air flow rather than the more typical low-velocity/high-volume air flow. Every time I mentioned this to one of the vendors, though, I was told "it doesn't exist" or "it doesn't work." Do you know who might manufacture such a system?

On a more general matter: I will require a 5-ton compressor, and the units being offered to me have a SEER (seasonal energy efficiency ratio) of 10.0 or 10.5. Are you aware of units in this size range with better SEERs?

—Thomas Darold, Pound Ridge, N. Y.

Jay Stein of Fine Heating Design in Denver, Colo., replies: I read your letter with some amusement. My immediate reaction was "Sure, I've heard of air-conditioning systems like that, but only for large commercial buildings." Just as I was about to sit down and write you a discouraging letter, I decided to do a bit of research first.

It turns out that there are two manufacturers of central air-conditioning systems for residential applications that use 2-in. I. D. high velocity flexible ducts. One is Hydrotherm, Inc. (Rockland Ave., Northvale, N. J. 07647; 201-768-5500). The other is Unico, Inc. (P. O. Box 144, Mt. Crawford, Va. 22841; 703-434-1081). The Hydrotherm system is marketed under the trade name "Space Pak." Both manufacturers have 5-ton units. I suggest you check with them to determine if there is an authorized dealer in your area. Because neither system is marketed in Colorado, I can't give you the benefit of any personal experience.

As for your second question, the SEER rating designates the amount of cooling a central air-conditioning system will provide per unit of electrical input. Lennox Industries, Inc. (200 South 12th Ave., Marshalltown, Iowa 50158; 515-754-4011), manufactures an HS14 Power Saver series with 5-ton units that feature SEERs of 12 and higher. These units contain two-speed compressors that operate at low speed when air-conditioning requirements are medium to light.

WATER-HEATER ANODE RODS

How do I know when the magnesium anode of a gas hot-water tank needs replacing?

—J. Kaye, Uniondale, N. Y.

Jay Stein of Fine Heating Design in Denver, Colo., replies: Water-heater anodes are often called sacrificial anodes for good reason. The glass lining that protects the steel tanks of most water heaters is not perfect. Electrolytic reactions with minerals in the water can cause the steel tank to corrode. The anode rod, which is usually made from magnesium, is the least noble metal in the tank. As a result, it's the first to corrode, sparing the steel tank from a similar fate.

Your water heater's anode should be replaced when it is substantially depleted. It should be checked every six months when other routine water-heater maintenance tasks are performed. In some heaters the anode rod can be accessed through a port in the top of the tank. Other manufacturers suspend the anode rod from the hot water outlet. Check the manufacturer's instructions for complete details.

You may be interested to learn that anode rods are not always made from magnesium. In some areas, the water contains minerals that release foul-smelling gases when they react with magnesium. Usually, it's sulfur that's most troublesome. In these areas anode rods made of aluminum are substituted. If you experience this problem check with the heater's manufacturer to obtain the correct anode rod.

INDIRECT WATER HEATERS

I have a five-year-old 150,000 Btu gas-fired hot-water boiler, which I am told is generous for heating my average-size home. I'm considering replacing my electric water heater because electric rates are high here. Also, we are planning a bathroom addition that will include a large whirlpool tub, so an adequate supply of hot water is important. In his article on automated home systems in *FHB* #53 (pp. 72-76), Jay Stein suggests replacing a traditional water heater with an integrated heater coupled to an ordinary boiler. Can you tell me more about the practical and financial considerations for converting an existing system?

—*Terrence W. Reigel, Warren, N. J.*

Author Jay Stein replies: Adding integrated hot water to your existing gas-fired heating system is an excellent idea. You need an indirect hot-water heater, which consists of an insulated 40-gal. storage tank and a built-in exchanger coil. My favorite unit is manufactured by Amtrol Inc. (1400 Division Rd. West Warwick, R. I. 02893), but there are others that many consider just as good. Most plumbing-supply houses carry at least one brand. A qualified plumber or mechanic can add one to your system provided you have a 2-ft. sq. space near your boiler.

To illustrate the economics of integrated water heating, let's start by calculating how many Btus are required annually to heat water for a family of four:

$$(4 \text{ people}) \times (20 \text{ gal./day}) = (80 \text{ gal./day})$$

$$(80 \text{ gal./day}) \times (90°F \text{ rise}) \times (8.3 \text{ Btu/gal. per } F°)$$
$$\times (365 \text{ days/yr.}) = (21.8 \text{ million Btus per yr.})$$

In the formula above: (90°F rise) is derived from the fact that water enters the house at 50°F and is heated to 140°F; 8.3 is the number of Btus needed to raise the temperature of 1 gal. of water 1°F.

I don't know what the utility rates are in New Jersey, but I imagine they're similar to what we pay in Denver. Heating water with a 90% efficient electric heater costs:

$$\frac{(21.8 \text{ million Btu/yr.}) \times (\$.076/kwh)}{(3,415 \text{ Btu/kwh}) \times (.90)} = \$539/yr.$$

Heating that same water with a 70% efficient indirect water heater integrated into a gas-fired heating system costs:

$$\frac{(21.8 \text{ million Btu/yr}) \times (\$.43/\text{ccf})}{(83,100 \text{ Btu/ccf}) \times (.70)} = \$161/\text{yr}.$$

In the formulas above, 3,415 is the number of Btus generated per kilowatt hour of electricity (kwh), and 83,100 is the number of Btus generated per 100 cu. ft. of gas (ccf).

Switching to gas saves $378 per year. In the Denver area you could have an indirect heater installed for under $1,000. That works out to be a payback period of less than three years.

You mentioned in your letter that you're adding a large whirlpool tub, but you didn't say how many gallons it would hold. With your boiler, a 40-gal. indirect heater could fill an 80-gal. tub in about ten minutes. If you need more hot water, you could add a second unit in parallel.

You should have the plumbing contractor install the indirect water heater on priority by using an additional circulator, two check valves and a relay. The heater should come with instructions on how to do this. When your boiler is heating water, the relay will prevent your main pump from circulating water through your radiators or baseboard heaters. Don't be concerned that your house will get cold while the tub is filling. The hot water sitting in your radiators or baseboard heaters will carry your home's heating load for this brief period of time. If you don't install the indirect heater on priority, the water heater and heating units will compete for energy from the boiler, and neither will do a good job.

MORE ON CRAWL-SPACE MOISTURE

Additional vents are often the solution to damp crawl spaces where we sometimes find termite, fungus and powderpost-beetle damage. However, in some situations (a site with a very high water table or a house built over a creek bed, for example) passive vents are not adequate during our six-month rainy season, and the crawl space remains so damp that I'm sure our repairs are only momentarily interrupting the damage/repair cycle. Do you know of a mechanical system that would automatically ventilate the crawl space and maintain the moisture content of the wood below 10%?

—Doug Carver, Oakland, Calif.

Charles A. Lane, a professional engineer in Shoreview, Minn.,
replies: In general, four factors influence moisture transport within
a crawl space: 1) the strength of the moisture source—the
incidence of moisture from plumbing leads, inadequate surface
and service drainage, high water tables or seasonal moisture;
2) temperature differences—the reaction of moisture vapor in
contact with a surface having a lower temperature; 3) moisture
transfer rate—the rate at which moisture is transferred from the soil
to the crawl space via hydrostatic pressure, capillarity, air movement
or vapor diffusion; and 4) circulation/ventilation rate—the rate at
which outside air is exchanged with the air in the crawl space.

You're correct that passive ventilation alone (i.e., open vents in
the crawl-space walls) may not be sufficient in crawl spaces over
very moist soil. In such cases I recommend taking steps to
minimize the moisture transfer rate before attempting to
mechanically ventilate the crawl space (assuming that steps have
already been taken to provide proper drainage and proper passive
ventilation). To minimize the moisture transfer rate, a continuous
ground membrane should be placed over the soil in the crawl
space, and all seams should be sealed with a flexible caulk. The
ground cover should have a perm rating of no more than 1.0 and
should be rugged enough to withstand foot and knee traffic; I
recommend 6-mil polyethylene.

All debris must be removed from the soil, and the soil must be
leveled before laying the ground cover. The edges should be
lapped at least 6 in. up the inside walls of the crawl space. The
ground cover should be sealed to the interior of the crawl-space
walls using flexible caulk or batten boards. It may also need to be
weighted down with gravel to resist the upward lift of soil gases that
are emitted from the soil.

At this time, very little technical information is available
regarding the relationship between the air humidity and the
moisture content of the wood in crawl spaces. Also, I am unaware
of any packaged mechanical control system that can regulate the
ventilation rates in a crawl space and maintain a 10% moisture level
in the wood materials.

An excellent source of information on controlling moisture in
crawl spaces is *Condensation and Related Moisture Problems in the
Home.* The manual costs $7.50 and is available from the Small
Homes Council-Building Research Council of the University of
Illinois at Urbana-Champaign, One East Saint Mary's Rd.,
Champaign, Ill. 61820.

DOWNDRAFT PROBLEM

I have a downdraft problem with my gas-fired, on-demand water heater that's located in the basement. The heater requires a 5-in. exhaust pipe. I used double-wall pipe, run on the exterior of the house and through the roof overhang. The vent ends about 2 ft. above the roof. The downdraft problem occurs on the coldest and calmest winter evenings. So much cold air rushed down one evening that it actually froze water in the heater's coils. Because this is an on-demand heater, it isn't running at all during the night, except for the pilot light.

Our basement is unheated and hovers around 50°F. In addition to our furnace, we also heat with wood and have a fairly tight home so the stove's air demand may contribute to the problem (although I closed the stove down several cold nights and this had no effect on the air rushing down the vent). How do I solve the problem?

—*Bob Eichenberg, New Marshfield, Ohio*

Jay Stein of Fine Heating Design in Denver, Colo. replies:
Downdrafting will occur when there is a negative pressure in the basement. This negative pressure is no doubt caused by your furnace and woodstove. Even when they are not operating, warm house air in their vents will cause what is called the chimney effect. House air will be drawn into the flue, rise up the vent and be expelled out of the house. Since the water-heater vent is run outside, on cold nights no chimney effect is developed because there is no warm air in the vent. It provides an ideal source to replenish the warm air leaving the house through the furnace and woodstove vent pipes.

You can relieve the negative pressure in the basement by providing outdoor combustion air to your furnace and woodstove. This is usually accomplished by running a duct from the outside to a point near the burners or firebox. Natural-gas appliances require at least one sq. in. of free vent area for every 5,000 Btu input. For a thorough treatment of venting techniques recommended by the National Fuel Gas Code, I suggest you get a copy of "Introducing Supplemental Combustion Air" (#DOE/CE/15095-7). This publication is available from the U. S. Department of Energy, Washington, D. C. 20585. For woodstoves there is no universally recognized code that's applicable. Although 4-in. round ducts are frequently used for this application, I would strongly suggest you check with the manufacturer as well as your local building code.

And finally, vent pipes for intermittently operating gas appliances should not be run outside in cold climates because flue gases tend to condense in them and corrode the pipe. Building an insulated chimney around the pipe, with a minimum of 1-in. clearance to the vent pipe, assuming it is Type-B vent, will be an improvement. Another technique that is popular with retrofitted instantaneous water heaters is to vent them directly through a sidewall with an induced-draft fan. Simply put, a small fan draws in flue gases and forces them outside through a vent hood. Tjernlund Products Inc. (1601 9th St., White Bear Lake, Minn. 55110) manufactures the fans, vent hoods and a special control kit for instantaneous water heaters. Best of all, no chimney is required.

VAPOR-BARRIER PLACEMENT

Is there an effective way of applying a vapor barrier in a two-story house where second-floor joists intersect with the roof's rafters over the wall (shown in the drawing below)? We are building such a structure and so far have applied the vapor barrier down the rafters and up the first-floor walls and part way across the ceiling. The R-19 fiberglass insulation in the roof extends over the wall and there is a soffit-to-ridge ventilation space above it, but this envelope cannot be covered with a continuous vapor barrier because of the location of the floor joists. Furthermore, the kneewall built upstairs would seem to keep this vulnerable

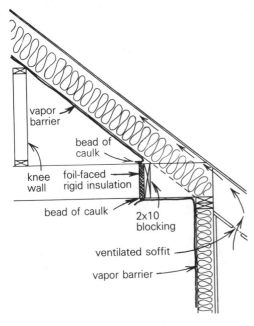

**area cool and encourage condensation. We have not
applied any wall or ceiling covering on the first floor, and
have noticed considerable condensation on the surface of
the vapor barrier on the first-floor ceiling and on the 2x10
blocking. How can this condensation be avoided?**

—Cherlynn Risch, Healy, Alaska

Gerry Copeland, an architect and builder in Spokane, Wash., replies:
The way to provide a continuous vapor barrier in this situation is to
cut ½-in. or 1-in. thick foil-faced rigid insulation to fit snugly
between the joists right against the 2x10 blocking. Rigid insulation
is easier to work with and can be sealed more effectively than if
you were to use small pieces of polyethylene between the joists.
Cut it exactly flush at the tops and bottoms of the joists. Bring the
vapor barrier up the first-floor wall, turn it along the ceiling joists
and seal it to the line of rigid insulation with acoustical (non-
hardening) caulk. Also caulk the joints on both sides where the
pieces of insulation meet the joists. On the second floor, bring the
vapor barrier down the rafters and seal it to the rigid insulation the
same way. This will provide the continuous vapor barrier that is
needed here.

The condensation that you noticed was probably caused by cold
air from the soffit vent leaking through gaps in the insulation batts
above the walls. Just a small amount of air passing here would
cause an untempered meeting of cold outside with warm inside air
at the vapor barrier.

One way to ensure complete filling of voids between framing
members is to use blown-in insulation, maintaining ventilation
space at the eaves with cardboard baffles. Also, your condensation
problem may be aggravated by inadequate insulation. You
mention R-19 insulation for a structure in Alaska. Here in eastern
Washington state, R-30 is considered a minimum.

VENTING WALLS

While researching superinsulated building techniques, I noticed that everyone stresses the importance of adequate attic ventilation. But no one recommends ventilation for side walls. One designer did mention that plywood or foil-faced insulation boards commonly used as exterior sheathing are impervious to the migration of water vapor through the wall. This could cause moisture to be trapped in the walls.

I am considering using a double-wall building technique that would provide an air space between the insulation and sheathing, as shown in the drawing below. Would this kind of system be superior to standard wall construction? Can you recommend any improvements?

—Mark Helling, Cincinnati, Ohio

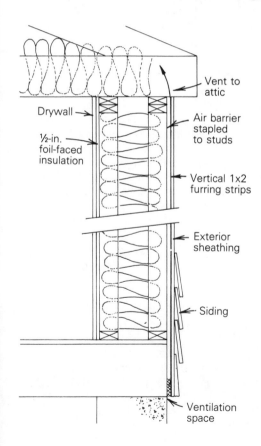

Vent to attic

Drywall

Air barrier stapled to studs

½-in. foil-faced insulation

Vertical 1x2 furring strips

Exterior sheathing

Siding

Ventilation space

John Hughes of Passive Solar Designs Ltd. in Alberta, Canada.
replies: The wall, as drawn, would be more than adequate
considering the climate in Cincinnati. By making a few changes
though, you can improve its energy efficiency and even lower the
cost (drawing below).

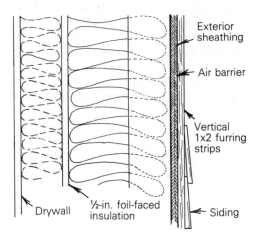

Exterior
sheathing

Air barrier

Vertical
1x2 furring
strips

Drywall

½-in. foil-faced
insulation

Siding

You've created a "vented rain screen" by strapping the wall with
vertical 1x2s before putting on the siding. This is an excellent way
of stopping rain from getting through the siding and into the wall,
should there be any holes in the siding. However, I would suggest
putting the sheathing directly on the studs, then the air barrier, the
1x2s, and finally the siding. There are three reasons for this. First,
the sheathing will provide greater racking strength if it's attached
directly to the studs. Second, the air barrier alone might not be able
to resist the pressure of the fiberglass batts. Since batts are meant to
completely fill a wall cavity, they're slightly oversized, and this
extra thickness could push the air barrier outward, filling the
ventilation space. And third, the sheathing will protect the
insulation from wind-washing, which will occur as air moves
through the air space. Wind-washing decreases the effective
R-value of the wall, increasing heat loss.

Sealing the foil-faced insulation will create an excellent vapor
barrier, as well as an air barrier, but you should move it to the
outside of the inner 2x4 wall. Then all the electrical wires can be
run in the inner wall, and there's no need to seal the boxes.
However, using the foil-faced foam simply because of the foil is an
expensive way to achieve a rigid vapor barrier. Since you already
have plenty of R-value in the wall, you might consider using foil-
faced hardboard, which costs less and would still serve as an
effective air/vapor barrier.

By building the inner wall first, you will be able to get at the outer face to seal the rigid insulation. Be sure to seal the insulation to the top and bottom plates with a bead of caulk. There's at least twice as much insulation on the cold side of the vapor barrier as there is on the warm side, so no condensation should occur on the foil.

If walls were vented the same way attics are—with the insulation exposed in the vent space—it would reduce the effective R-value of the insulation. Air would enter at the bottom of the wall, be warmed by heat loss from the house, rise and exit into the attic. In doing so, it would draw more air into the wall cavity, robbing more heat from the house. Attics are vented to carry off moisture escaping from the house, but the air enters the attic above the top of the insulation and doesn't pass through it. As the outside air temperature drops, the effectiveness of attic ventilation is reduced, since cold air can't carry off as much moisture. If you find frost in the attic, don't increase the attic ventilation—stop the flow of moisture from the house to the attic by tightening the air/vapor barrier.

ADA OVER RIGID INSULATION

I'd like to use the airtight drywall approach _(FHB_ #37) in remodeling my house. The walls are 2x4 construction, and I'd like to add rigid insulation on the inside. What type of insulation should I use? Should I do anything different with the caulking back?

—_Jeffrey L. Ellis, Meadville, Pa._

Architect Tom Bowerman replies: I suggest using polyurethene foam, because of its high R-value per inch. This permits interior application without losing much interior space. I'd use 1½-in. foam and fur out the top and bottom plates with 1½-in. spacers to give the drywall a firm backing. Fire-rated drywall is probably required in your area, because foam insulation is flammable.

The backer rod would be used in the same manner as in new construction, and the furring strips would be gasketed to the floor as well as to the drywall. It might be a good idea to use vertical furring strips on the studs and install the foam between them. But you could avoid this by using long drywall screws that would pull both the drywall and the insulation up snug to the studs, eliminating the possibility of movement and joint cracking. With the latter method, I'd definitely use fiberglass-mesh joint tape.

RETROFITTING VAPOR BARRIERS

My wife and I recently bought an 1854 grist mill in northeast Maryland. It's a chestnut post-and-beam structure, with pegged mortise and tenons. The building has clapboard siding (¾ in. by 8 in.) and no interior walls. We plan to leave the posts and beams exposed and drywall or plaster between them (posts are about 9 in. square and 10 ft. o. c.). This means that the wall insulation and vapor barrier will be between the posts, which will then be a thermal break.

Will this cause a great risk of condensation on or in the posts (many are severely checked)? Should we consider instead temporarily removing the siding to install a continuous vapor barrier and rigid insulation? We would prefer not to disturb the good parts of the siding, which is original to this historic building, but also do not want to risk damage to the structure.

—William G. Shimek, Darlington, Md.

Architect and designer Larry Medinger of Ashland, Ore., replies: Damaging water deposits result from warm, moist interior air coming in contact with a cold surface. Air temperature near such a surface drops below the dew point, and water vapor in the air condenses on the surface. In conventionally framed walls, air/vapor barriers are installed to prevent interior air from being driven or sucked into wall cavities or attics through openings in the interior wall surface. This interior air may reach areas in or beyond the wall insulation where surface temperatures are below the dew point.

Hardwoods have an R-value of approximately .9 per inch, depending on species and moisture content. At their full 9-in. thickness, your uninsulated beams would be rated at about R-8. Deep checks between interior and exterior surfaces can lessen R-values. At such low insulative levels, interior surface temperatures can reach dew point during colder weather.

If you decide to insulate and add a vapor barrier between the posts, you could use two strategies to protect your beams from condensation. First, maintain low humidity levels in the home by modifying living habits (i.e., taking shorter, cooler showers); by installing bath and kitchen vent fans on timer switches; by installing a humidistat controlled air-to-air heat exchanger; and by installing a dehumidifier.

Second, to prevent water absorption, apply a good surface sealer to the posts, taking care that coverage is complete deep within the cracks in the wood. It is not necessary to use a treatment with fungicide or other poison in it. Be sure to seal the joints between the air/vapor barrier in the adjoining wall segments and the sealed surface of each beam. A problem with this method is that untidy surface condensation may still collect on the sealed surfaces. A combination of the above solutions may be best.

The adaptation of historic structures to modern uses is often expensive, and you always run the risk of destroying what you are trying to preserve. However, if done correctly, the alternative of removing the siding and installing rigid-foam insulation is a sound one that will yield the most completely weatherized home. Such insulation generally will be rated R-5 to R-7.5 per inch of thickness, depending on type of material and manufacturing process. I recommend using a urethane or phenolic foam board that is at least 1½ in. thick (about R-10). At this thickness, 2x2s are handy to fur out door and window jambs, and can be used as nailers wherever it is necessary. With the added exterior insulation, your interior insulated wall segments can be of narrower stock (R-11). This will save you some money and result in more of your beam's profile being visible.

ALGAE-STAINED SHINGLES

Two years ago I replaced my roof with top-quality shingles. Now I have a streak of green moss on the roof, parallel to the dormer. What does this indicate? The roof is fully insulated on the underside, and the house is located next to a pond.

—R. Frost, Georgetown, Mass.

Todd Smith of Rogers Roofing in Verona, N. J., replies: The discoloration on your shingles is caused by algae referred to as "fungus growth," resulting from wind-blown spores that landed on the roof and began to grow. The insulation is probably not related to the algae growth, but the pond very well may be. The pond creates a humid environment in which algae thrive. You don't say whether the stains are on the sunny or shady side of the roof, but I assume they're on the shady side because the sun would help deter the fungus growth.

One way to correct this condition is to clean the roof with a chlorine-bleach solution. You can make your own with household bleach or use a commercial cleaner made for this particular problem.

The roof should not be scrubbed because this will remove the granules. Cleaning with a sweep broom or whisk broom is advisable. Be careful to avoid damaging any shrubbery when using and rinsing the solution. A roof treated in this manner should stay clean for approximately six months.

Another way to remove the algae is to install a piece of galvanized metal under the shingles above the stain. The metal should be installed with approximately 1 in. to 2 in. exposed. In time, the rainwater that runs over the galvanized metal and onto the stained shingles will clean the shingles, and prevent further algae problems.

Some shingle manufacturers make a stain-resistant shingle to prevent algae problems. Stain-resistant shingles contain a percentage of zinc granules that inhibit algae growth.

LONG-HORN BEETLES

We are apparently sharing our log home built of old timbers with long-horn beetles, which we have identified as the ivory-marked beetle *(Euburia quadrigeminata).* During construction in 1982, we had an exterminator treat the ground and the timbers, but the rest of the lumber arrived after he left. As a result, our oak subflooring and replaned oak-finish flooring, reclaimed from a demolished factory, as well as the cherry window and door framing, went untreated and now show occasional signs of this pest. We find from 4 to 6 adults, presumably dying, wandering about the floors in late spring each year, and not necessarily near where we see the tiny piles of sawdust around new holes that are about ⅛-in. dia. We also have seen sawdust in the fiberglass batts of insulation and traces of borer-like channels on the underside of the subflooring.

We hesitate to call the original exterminators because they are a small-town franchise operation that has changed hands twice since we used them. While they may have licenses and equipment to buy, sell and apply the chemicals, we don't think their knowledge and experience goes much beyond mice, termites and fleas.

We aren't concerned with a couple of bugs and holes each year, but if the beetles are causing serious damage, we want to get rid of them. Any suggestions?

—*Jack Remington, Vevay, Ind.*

Phil Pellitteri of the Insect Diagnostic Lab at the Cooperative Extension Service of the University of Wisconsin, replies: The ivory-marked beetle *(Euburia quadrigeminata)* is a type of long-horned wood-boring beetle. It is a common eastern species that is most often associated with ash, hickory or locust. It attacks dead and dying trees and normally requires two years to complete its life cycle, but this species has been found emerging from flooring, sills and other wood 10 to 15 years after the material has been installed. The important point is that it cannot re-infest structural wood. Any beetle emerging was in the wood when the house was built. This insect is most commonly brought into a home via firewood; adults are also attracted by nightlights.

From the background you've given me I don't think this is the insect that's causing the holes or the sawdust. This long-horned beetle is a rather large 1-in. long insect that would leave irregular holes about the diameter of a pencil (or bigger) and a rather coarse sawdust behind.

I think your house may be infested with some type of powderpost beetle. The age and history of the wood, small fine sawdust and small emergence holes are characteristic of these insects. Powderpost beetles can become a structural problem in some situations, depending on the starch and moisture content of the wood. They are slow-working insects that prefer unfinished or roughsawn wood.

There are a number of treatments, such as wood replacement, altering the moisture problem, chemical sprays and fumigation, but the cost of control versus replacement, the severity of the damage and the potential for further activity need to be assessed. It may even be that no controls are needed. I would suggest trying a different pest-control firm that has a better handle on the problem. I would not spray or fumigate unless absolutely necessary. The treatments that were used during construction will not be effective for this problem.

LAMINATING STRUCTURAL TIMBERS

I am building a post-and-beam house and I need information on how to make "built-up" posts and beams. I want to construct posts of three or four 2x6s, and beams of three 2x10s. For esthetic reasons, I want to avoid a lot of nailing, so I'm wondering whether any adhesives are available that would do the job just as well. I've used epoxy for a couple of boats I've built, but that could be awkward for my house—many of my posts and girders will have to be assembled in place, and clamping lumber together would be difficult. What about construction adhesives such as Lumberlock? Your article on construction adhesives *(FHB #58, pp. 72-75)* had little or nothing to say about such problems.

—*Donald H. Ericksen, Huntsville, Ariz.*

Stephen Smulski, Assistant Professor of Wood Science at the University of Massachusetts at Amherst, replies: Construction adhesives are inappropriate for laminating dimension lumber. Construction adhesives are actually classified as semi-structural because of the uncertainty surrounding their load-bearing capacity and long-term performance. Construction adhesives can soften and weaken at the elevated temperatures found in attics and south-facing walls, can be dislodged by water, and can creep (stretch out) under sustained loads. In fact the major building codes recognize that nailed-glued floor systems, for example, are stiffer, but no stronger than nailed-only floors.

If you want fewer fasteners to show, your best bet would be to use solid sawn timbers or commercial glue-laminated timbers. Commercial glulams are made with prime structural phenol-resorcinol adhesives, which aren't available over the counter. Other options would be to nail-laminate members, then box them in with finish lumber, or to use fasteners with architectural heads designed to show.

As long as stress-rated dimension lumber is used, even a nail-laminated beam is stiffer and stronger than a solid sawn timber of the same species and grade. This is because knots, cross grain, and other strength-reducing defects are confined within each lamination. When making laminated timbers, remember that the nail type, nail size and nailing pattern are crucial. Using too many nails too close together in a straight line will encourage along-the-grain splitting, which weakens members.

The trickiest part of nail-laminating is splicing individual members to get longer lengths. Avoid splices whenever possible; even when done properly, they are still the weak link in nail-laminated members. But splices are often unavoidable, so here is a

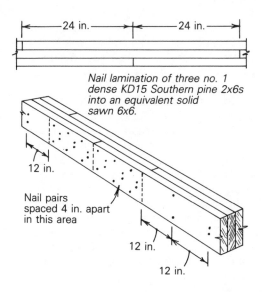

Nail lamination of three no. 1
dense KD15 Southern pine 2x6s
into an equivalent solid
sawn 6x6.

24 in. — 24 in.

12 in.

Nail pairs
spaced 4 in. apart
in this area

12 in.

12 in.

plan for nail-laminating three #1 dense KD15 (15% moisture content) Southern pine 2x6s into the equivalent of a solid sawn 6x6 (drawing above). Threaded nails, 3 in. long and .120 in. dia., are driven through both faces into the center laminate, with the two nailing patterns slightly offset.

TUDOR TIMBER FRAMES

I plan to build a Tudor-style home using a real timber frame with single-brick inlay between the timbers. Because the timber frame will be exposed to a rainy West Coast climate, would a Douglas-fir or spruce frame be susceptible to premature rot? Should I consider using cedar instead? Also, if I use green timbers, will the drying and shrinkage movement of the timbers cause much damage to the bricks and mortar?

—Ross Walton, Denman Island, B. C.

Charles Landau of Timbercraft Homes in Port Townsend, Wash., replies: Building a Tudor or half-timbered house presents three major problems: shrinkage, timber rot and insulation. Timbers do most of their shrinkage across the grain rather than along their length. This means that if you build your frame in such a way that the posts run all the way through from sill to rafter, then the timbers will shrink away from the brick. Using this system necessitates an ongoing program of caulking until the timbers have reached their maximum shrinkage.

The problem of rot is closely related to that of insulation. If the house were not insulated and heated on the inside, the timbers would have much less tendency to rot. Insulating and heating the interior space creates a temperature gradient between the inner and outer skin. Because of this, moist warm air from the inside will be cooled to its dew point somewhere inside the wall cavity. The resulting condensation will eventually cause the timbers to rot.

The best solution would be to enclose and protect the timber-frame with stress-skin panels. To give the desired Tudor look, 4-in. timbers could be applied to the outside of the panels, then infilled with brick. The surface of the brick should be flush with the outside of the timbers. The top surfaces of these timbers should be beveled down and away from the stress-skin panels to promote better drainage. Using dry wood for this part of the project will help reduce the amount of caulking needed in the future. Although this would be an expensive system, it would last much longer than the one you describe. As for what wood to use, cedar would certainly be a good choice for the outside portion and Douglas fir for the structural frame.

SEALING CREOSOTE BEAMS

The ceiling of my house consists of 12 exposed oak beams measuring 15¼ in. by 7 in. and spanning the length of the house. The beams are old narrow-gauge railroad timbers that were impregnated with creosote as a perservative. The house is only six years old, but the beams range in age from 20 years to 90 years. Supposedly they were sandblasted when installed.

When I purchased the house five years ago, I had the beams scraped, wire brushed, washed with naphtha and sealed with three coats of a polyurethane-type sealant. This reduced the creosote odor by about 80%. But the coating has apparently been decomposing because the odor is increasing again.

Do you know of a sealant that will eliminate the creosote odors? Incidentally, the FDA has banned creosote for consumer use.

—N. M. Sullivan, Santa Fe, N. Mex.

Terry Amburgey, an associate professor with the Forest Products Utilization Laboratory at Mississippi State University, replies: At the outset, I must emphasize that wood treated with oil-borne preservative chemicals that have a high vapor pressure (such as creosote or pentachlorophenol) should not be used in habitable spaces. Volatile components of these materials will continue to migrate to the wood surface. When wood treated with these chemicals—especially freshly treated wood—is exposed in a poorly ventilated area, the vapor level can become sufficiently concentrated to cause irritation of mucous membranes as well as other discomforts.

There is no known way to contain creosote odors indefinitely. The only permanent solution to your problem would be to replace the creosote-soaked beams with untreated ones. As you have discovered, however, coating creosote-treated wood can temporarily retard the movement of volatile components. Studies have shown that polyurethane treatment can decrease the creosote odors by 70% to 90%. If you do decide to recoat, I recommend that you apply two coats of polyurethane and increase the house ventilation.

EVALUATING DECK WOODS

Having lived with an outdoor deck of standard-grade 2x4 fir and watched it splinter, crack and check over the years, I've decided to install a deck where cost is no object. I'm considering three different types of wood—vertical-grain fir, vertical-grain cedar and vertical-grain redwood. Which will last longest, and splinter and crack the least? I'm considering using bleaching oil to give the deck a silvery grey color. What subsequent maintenance is suggested? The deck will be on Long Island, mostly in full sun, and will encircle a swimming pool.

—P. J. Mejean, New York, N. Y.

Scott Grove, a builder in Rochester, N. Y., replies: It's not often that I get a chance to spec out a cost-is-no-object deck, but I have pondered the possibilities many times. As for which lumber will last the longest, I would vote for redwood for the following reasons.

Deterioration is my first concern. When comparing deck woods, I look for the highest natural resistance to decay. Douglas fir has moderate resistance; redwood and cedar are rated as having high decay resistance.

Stability, which will help control warping and checking, is my second concern. The radial coefficient—the percentage a board shrinks perpendicular to the growth rings from green lumber to kiln dry at 6% moisture content—is the best shrinkage indicator for vertical-grain lumber. The radial coefficient of fir is roughly 4.5%. Western red cedar and redwood are both around 2.5%.

The reason I would choose redwood over cedar is density. The average relative density for cedar is .32, while that for redwood is .38 (relative density is the ratio of a wood's moisture-free density to the density of pure water at 4°C). Given an equal amount of traffic, this means less wear on a redwood deck than on a cedar one. You should also be certain to lay the decking with the bark side up. The denser, harder bark side will resist wear better.

Of course, some people will see redwood's density as a disadvantage, because in general, the denser the wood, the more it will shrink and swell. But the numbers notwithstanding, I don't believe there is a significant difference between redwood and cedar, although I'm sure there are many opinions on this subject.

By special ordering vertical-grain redwood, you can't go wrong. Vertical-grain boards are inherently less prone to deformation than other boards of the same species. The relatively straight, parallel and short growth rings ensure a stable board. When placing your order, specify heartwood only (I'm keeping in mind that cost is no object). The heartwood has a much greater resistance to decay than the sapwood. Note, however, that special ordering will require a knowledgeable contractor and a patient lumber dealer. You can't just call in the order for delivery the following week.

You can also request that the lumber be kiln-dried to the average moisture content for your region, which is around 19% for New York and New England. Kiln-drying will help reduce the swelling and shrinking that wood undergoes in adjusting to the local environment. This process, along with improper installation, accounts for the majority of cracks and checks.

Regardless of the wood you choose, an ongoing maintenance schedule is crucial. Because the market for deck coatings is relatively new and always changing, however, choosing a preservative, repellent or stain can be confusing. If your decking has a high natural preservative content, like redwood, your main concern will be how to fight mildew and repel water.

Water repellents come in a variety of formulas, and you can also buy deck stains that contain a water repellent. Labels on the more effective products should state that they meet or exceed the Federal Specification TT-W-572B. The product label should also indicate that it resists mildew.

There is an easy test to determine when you can apply a finish. Sprinkle a few drops of water onto the board to be treated. If the water is quickly absorbed into the board or if it wets out quickly, the board is ready to receive a finish. Stay away from any film-forming finish that will trap moisture in the wood and do more harm than good. The finish must let the wood equilibrate (breathe or pass vapor), but must also inhibit liquid from entering the board.

Proper application techniques are essential for a long-lasting finish. Follow the manufacturer's recommendations. It is highly advisable to add additional preservative to the end grain of a freshly cut board and into any drilled holes. Before applying any type of finish, I recommend that you clean the deck with a solution of three parts water and one part bleach, followed by a detergent or trisodium phosphate solution. This procedure will help to remove fungi, mold and algae from the surface.

As far as bleaching the deck grey, you may be rushing it a bit. Ultraviolet radiation will quickly turn the deck silvery grey on its own. Use a clear repellent. Then let the deck turn grey by itself.

One last note. Expensive lumber will not by itself prevent warps, splinters and checks. It is essential that the deck be designed and constructed with proper techniques (for a complete discussion of deck design and construction, see *FHB* #29, pp. 42-46).

HARVESTING YOUR OWN LUMBER

I have a chance this fall to cut some yellow poplar off a strip of land being cleared, and I plan to use this wood to build a house. Where can I get information regarding allowable spans for poplar? Also, if I air dry the lumber, how would it compare with kiln-dried lumber?

—Zestrell A. Turner, Rockvale, Tenn.

Christopher F. DeBlois, a structural engineer with Palmer Engineering in Atlanta, Ga., replies: Whether or not span tables are handy, you can always go back to basics and calculate the allowable loads and spans by the same method used to formulate the tables. The first required reference is *National Design Specification for Wood Construction* (NDS), published by the American Forest and Paper Association (1250 Connecticut Ave., N. W., Suite 320, Washington, D. C. 20036; 202-463-2700). The latest version contains all the formulas and design criteria that govern virtually every aspect of structural design with wood. In addition, *Design Values for Wood Construction,* a supplement to NDS, lists the basic allowable stresses for every major species, size and grade of dimension lumber. The set costs $25.

The other side of this business involves cutting, drying, finishing and grading the lumber. Once the lumber is rough cut, be sure it is thoroughly dried before milling it to finished dimensions. As for kiln drying vs. air drying, dry is dry. Kiln dried or air dried, the strength of the lumber will be the same for the same moisture content. However, drying lumber in a kiln kills bugs in the lumber; air drying doesn't. Either way, it is important that the lumber's moisture content be no more than 19% because the wetter the lumber, the lower its load-bearing capacity. As the lumber dries to below 19%, twisting, warping and cupping are likely to occur— better to let them happen in your drying stacks before finishing than in your new house afterward.

Finally, you must have the lumber graded. Section 1701.4.1 of the 1991 Standard Building Code and sections R-402.1, R-602.1 and R-702.1 of the 1989 CABO One and Two Family Dwelling Code require that any lumber used for load-supporting purposes be identified by the grade mark of an approved lumber grading or inspection agency or by a certificate of inspection providing the same information. Approved agencies are those certified by the American Lumber Standards Committee. Two agencies in your area that are likely to have either a mill where your lumber can be graded or a traveling representative who can grade the lumber at your site are the Southern Pine Inspection Bureau (4709 Scenic Highway, Pensacola, Fla. 32504; 904-434-2611) and Timber Products Inspection (P. O. Box 919, Conyers, Ga. 30207; 404-922-8000).

HANDLING PRESSURE-TREATED WOOD

My local lumberyard man refuses to handle pressure-treated material without wearing gloves because of the "dangerous chemicals" used in pressure-treating. Is this a health hazard for carpenters who handle and saw this material? What chemicals are used? Are gloves and dust masks adequate protection?

—Michael Gillogly, Oakland, Calif.

Stephen Smulski, Assistant Professor of Wood Science and Technology at the University of Massachusetts at Amherst, replies:
Most treated lumber and plywood sold by retailers is protected against decay with either CCA (chromated copper arsenate) or ACA (ammoniacal copper arsenate). Referred to as "pressure-treated" because of the process used to drive the preservative deep into the wood, the hallmark of these products is their green tinge. The arsenic component has piqued concerns about safety. One property of arsenical preservative that makes CCA- and ACA-treated products appropriate for consumer use is that both preservatives react chemically with the wood and become permanently fixed to it. This means that the preservatives become insoluble and cannot be readily leached from the wood. Because both are applied as a water solution, no vapors are later emitted.

Is the gloved yardman overreacting? Because of precise control of the treating process, virtually all properly treated wood is surface-clean. Infrequently though, a white bloom of residual preservative, formed through a phenomenon called "sludging," shows up. Because your yardman is likely handling a large number of pieces daily and can't possibly wash his hands each time he handles treated wood, he's taking a logical precaution.

Should builders and do-it-yourselfers be concerned? The two routes for arsenic entry into the body are inhalation and ingestion, especially of airborne sawdust. I recommend wearing a dust mask (and eye protection) when machining *any* wood product, whether treated or untreated. Though sometimes cumbersome, gloves will certainly prevent skin contact; washing one's hand after handling treated wood seems to work just as well.

A 1976 study of carpenters in Hawaii, where entire homes are built of treated wood, showed no difference between the health of builders who had left the trade before the introduction of treated wood and those who hadn't. The authors of a 1977 study of workers building foundation components from treated lumber and plywood in a Wisconsin manufactured-housing plant reached the same conclusion. Neither masks nor gloves were worn in either case.

An adage among toxicologists says, "The dose makes the poison." From everything I've gleaned from the scientific journals, the bottom line appears to be that even without masks or gloves, exposure to arsenic from handling or machining wood protected with CCA or ACA is well below the average 80 micrograms of arsenic Americans get daily from their food and water. For more information, ask your retail dealer for the EPA-approved Consumer Information Sheet for Inorganic Arsenical Pressure Treated Wood.

SEALING PRESSURE-TREATED LUMBER

Recently, a co-worker asked me a question I couldn't answer. He is building a deck with Wolmanized lumber and wanted to know if the life of the wood could be extended by sealing or otherwise treating the lumber with preservative. He has gotten conflicting answers from several people. Some say using a sealer would counteract the treatment already used on the lumber. Others say that sealing would help to extend the life of the wood. Can you settle the matter for us?

—*Howard Day, Philadelphia, Pa.*

Scott Grove of Effective Design Incorporated in Rochester, New York, replies: First, the pressure-treated wood generally used for deck building is most commonly preserved with chromated copper arsenate (CCA). This water-borne preservative reacts chemically with the cellulose in wood and locks itself in. It will not leach out, and adding any additional preservative will not counteract this process. During the pressure-treating process the preservative is "pushed into" the wood, though it does not always completely penetrate the wood, especially on larger timbers. Heartwood does not accept the preservative at all. Keeping this in mind, whenever cutting or drilling into a board, the end grain should be chemically preserved. End grain is most vulnerable because it acts like a straw to suck up moisture. When placing a post onto a footing, for example, place the factory end down if you can. It has about a 12-in. penetration of preservative along the end grain.

The wood-preservative industry does recommend applying a water repellent after the wood has had a chance to dry out. They say you should treat the wood one month after installation, but I recommend waiting at least three months. It all depends on the moisture content of the wood. In the New England area, pressure-treated lumber should stabilize at around 19% moisture content, and at that point the wood will accept the greatest amount of water repellent.

Pressure-treated lumber is normally kiln-dried before it is treated, and unless you specify KDAT (Kiln-Dried After Treatment) you will basically be purchasing wood with about 70% moisture content. When the wood reaches 30% moisture content, it will tend to warp, so be sure to keep the lumber stacked and bound until use, but that's another story.

MIXING CEMENT WITHOUT LIME

I am building a concrete-block house in the Caribbean. My concern is with the mortar I will be using. There is no lime available on the island. Everyone uses a mix of a mixture of No. 1 portland cement and fine sand. Will this mixture be strong enough, or should I plan on bringing my own supply of lime to add to it?

—*Charles Williams, Oakville, Ont.*

Builder Charles Wardell replies: According to a member of the technical staff at the Portland Cement Association (5420 Old Orchard Rd., Skokie, Ill. 60077-4321; 708-966-6200), a mixture of portland cement and sand is more than adequate for laying block. In fact, it may even be somewhat stronger than the standard water/sand/lime mix. That's because lime is added for convenience. Lime makes the wet mud more plastic and cohesive—and thus more workable—but it is by no means necessary. While a batch of mud with no lime may not be the easiest working stuff, structurally it's fine.

SPALLING CONCRETE

The poured-concrete basement walls of my 1930s home are crumbling. Sections of the wall simply break apart when touched. Some spots have crumbled to a depth of 2 in. What causes this to happen and how might I fix it?

—Tom Gordon, Springfield, Ohio

Construction consultant Alvin Sacks of Bethesda, Md., replies: The problem that you cite is usually referred to as spalling, popout or pitting. A few possible causes come to mind. If the walls consisted of low-quality, permeable concrete to begin with, then water may have migrated through them from the outside. If the surrounding soil, and thus the groundwater, is also acidic, they may have neutralized the alkaline concrete, making it more likely to crumble. If the basement or crawl space wasn't heated, the problem could have been caused by the freeze/thaw action of the migrating water.

For repairs to the surface, use a "top and bond" cement product. It won't be permanent, but it will be cosmetically acceptable. For a longer-lasting (and more expensive) repair, use an epoxy-concrete or polymer-modified concrete mixture.

STIFFENING A FLOOR FOR TILE

I want to lay 8-in. by 8-in. quarry tiles on a second-story loft floor. The 10-ft. by 20-ft. floor is framed with 2x10 joists laid 24 in. o. c. A layer of ¾-in. T&G plywood was laid perpendicular to the joists. A second layer of ¾-in. plywood was laid parallel to the joists. Both layers of plywood were fastened to the joists with screws.

I'm concerned that there's still too much bounce in the subfloor and that the tile might crack because of it. Should I abandon my idea of installing tile, or is there some way to stiffen the floor?

—Dari S. Hing, Scottsdale, Ariz.

Bob Wilcoxson of Island Tile in Edgartown, Mass., replies: You'll get a stiffer floor by installing both layers of plywood perpendicular to the joists (drawing on p. 196). It's important to stagger the joints and to leave a ⅛-in. gap between sheets. You'll add even more stiffness by gluing the sheets together with panel adhesive and driving 1¼-in. screws into them every 6 in. After the plywood is down, cover it with galvanized expanded-metal lath nailed every 6 in. with roofing nails. Lay your tiles in a thinset mortar mixed with a latex additive. The mortar should be designed for bonding to plywood

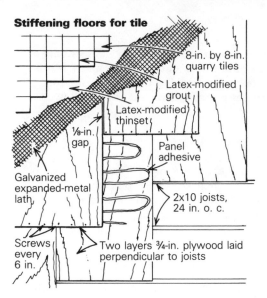

Stiffening floors for tile

8-in. by 8-in. quarry tiles

Latex-modified grout

Latex-modified thinset

⅛-in. gap

Panel adhesive

Galvanized expanded-metal lath

2x10 joists, 24 in. o. c.

Screws every 6 in.

Two layers ¾-in. plywood laid perpendicular to joists

and should be installed in two layers. Spread the first layer with a flat trowel, forcing the mortar into the lath and trowelling the surface flush with the top of the lath. You can lay the tile the next day with the same mix, using a ⅜-in. by ¼-in. notched trowel. Wait at least 24 hours, then grout the joints using a sanded grout with latex grout additive.

TILING OVER CRACKED CONCRETE

I put an addition on my house four years ago. Both the original house and the addition have concrete slabs. The slabs are covered by a ceramic-tile floor. Where the new slab meets the old, however, the concrete continues to crack from shrinkage. It cracks right through the 8x8 tiles. I have cleaned the crack and filled it with epoxy grout to no avail. I'm now considering removing the cracked tile and putting waxed paper under it. Any better ideas?

—Harry Rickard, Florence, Ore.

Bob Wilcoxson of Island Tile in Edgartown, Mass., replies: As I see it, you have two options. If you can live with the crack but don't want it to get worse, then the easiest route is to cut an expansion joint in the tile over the crack and fill it with a flexible caulk or an expansion-joint material such as premolded PVC. A more thorough solution is to remove a course of tile on each side of the crack and install a membrane (drawing, facing page). Instead of waxed paper, I recommend a crack-isolation membrane such as Nobleseal TS.

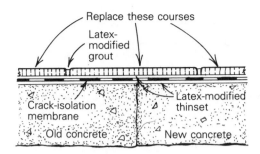

Replace these courses

Latex-modified grout

Crack-isolation membrane

Latex-modified thinset

Old concrete

New concrete

When removing the old tile, be sure to cut the joint between the course remaining and the one being removed. This decreases the chance that you'll chip the remaining tiles. Completely clean the concrete, then lay the membrane in a bed of latex-modified thinset. The tiles can be reapplied over the membrane using the same mix. With any luck, the membrane will allow the concrete some movement while keeping the tile above it in one piece. When regrouting, use a latex-modified grout.

A MATERIAL DIFFERENCE

Please explain the difference between interior and exterior materials. I recently enclosed a porch using rock lath and plaster for the inside and stucco for the outside. Now, faced with patching a plastered bathroom ceiling, I wonder: why not knock out all the old plaster and replace it with stucco? Wouldn't it stand up as well to water vapor as it does to rain? Then there's paint. I have been using exterior-grade paint for inside trim because my doors and windows are usually open year-round. Finally, what's the difference between interior- and exterior-grade drywall and plywood?

This all sounds pretty basic, but the only answer I ever get is that the exterior materials "hold up" better.

—Patricia Welch, Topanga, Calif.

Kevin Ireton, editor of Fine Homebuilding, *replies:* According to the Portland Cement Association (5420 Old Orchard Rd., Skokie, Ill. 60077; 708-966-6200), stucco and plaster are the same material: a mixture of portland cement, sand and water. The term "stucco" is usually used to describe plaster used outdoors, while "plaster" generally implies interior use.

As for paints, I was told by the National Paint and Coatings Association (1500 Rhode Island Ave. N.W., Washington D. C. 20005; 202-462-6272) that exterior paint is more durable when exposed to weather and that interior paint comes in a larger selection of colors. I also called the Forest Products Laboratory in Madison,

Wisconsin, and learned that one of the key differences betweeen interior and exterior paint is that exterior paints generally include a mildewcide. Beyond that, the difference is in the resins and pigments used in exterior paint to help it survive the expansion and contraction caused by temperature swings, as well as by the damaging effects of exposure to water and to the sun's ultraviolet rays. Because of all these extra things in exterior paints, you need to be careful about their potential toxicity when using them indoors.

The difference between interior and exterior plywood is mostly in the glue that's used. Exterior glue will not break down when exposed to water. Besides the difference in glues, exterior plywood must be composed of C-grade or better veneers, whereas interior plywood can use D-grade veneers.

There are two kinds of drywall used outdoors. One is used as a substrate for siding (like stucco). Its core is made water-resistant with asphalt emulsions; the multilayered face and back paper are made water-resistant with other chemicals. This is very similar to the treatment given to the moisture-resistant drywall used in bathrooms (whose green tint, by the way, has nothing to do with water-repelling chemicals; it's just used to distinguish moisture-resistant from standard drywall). The asphalt emulsions make the drywall a bit more limp than normal, so sheets treated with them are not recommended for overhead installations (when used this way they have a tendency to sag). The other type of exterior drywall is specifically designed for soffits and porch ceilings. It has the same core as standard drywall, but is wrapped with a special paper that seals it from moisture.

CONCRETE MASONRY BRICKS

I plan on building my own home and would like to use brick. However, I find the cost of brick to be prohibitively expensive when compared to vinyl siding or free fieldstone. Nonetheless, the weathering ability and general appearance of brick is highly desirable. As such, I am considering making my own cement brick, much like that used for patios. I plan to do this by pouring a cement and sand mixture into oiled, homemade forms. I suspect that by adding pigment I could choose any color.

My questions are as follows: Is the structural integrity of such brick equivalent to kiln-fired clays? Would such brick weather as well? What would you recommend as far as cement type, and the proportions of cement and sand? How would the cost compare to brick at 30¢ for the standard building brick?

—Michael S. Niziol, Apalachin, N. Y.

Jacob W. Ribar, principal masonry evaluation engineer for Construction Technology Laboratories, Inc. in Skokie, Ill., replies:
You need to make several decisions based on cost, personal labor, appearance, maintenance, and resale value. My choice is (1) fieldstone, and (2) clay units.

Comparing clay brick to vinyl siding is inappropriate unless you consider long-term benefits. The initial cost may favor siding, but long-term costs are definitely in favor of clay brick.

I don't know what kind of patio brick you have seen. There are high-strength concrete pavers on the market that are used not only on patios but also as highway pavements. These units are produced in various shapes to create patterns in the pavement. Though very dense and durable, these bricks are produced with specialized equipment that's not available to a "do-it-yourselfer."

The main concern with homemade bricks is durability, rather than strength. Some concrete products do not perform well when installed in a horizontal position in a freeze/thaw environment. Homemade units may have a similar problem even though they were installed on a vertical wall. Color consistency would also be a challenge.

You could consider buying concrete masonry bricks from your local concrete-block producer. They are about half the cost of clay units. You should be aware, however, that concrete masonry walls need to receive a coating (paint or a clear finish) to reduce water penetration. Thus, they pose a maintenance problem not found with clay units.

MASONRY BAKE OVENS

I would like to build a small brick oven in my basement and fuel it with my woodstove. However, I have tried unsuccessfully to find a mason, or anyone else for that matter, who knows how to build one. I am willing to do the work myself, but can't find any information about what is necessary.

—John Parziale, Melrose, Mass.

Albie Barden of Maine Wood Heat Co. Inc., replies: First, there are two books I'd suggest reading. One is *The Forgotten Art of Building and Using a Brick Bake Oven,* by Richard M. Bacon (Yankee, Inc., Dublin, N. H., 1977). It is probably out of print, but should be available from a good library. The other, which is still in print, is *The Bread Ovens of Quebec,* by Lise Boily and Jean Francois Blanchette (The Canadian Center for Folk Culture Studies, National Museum of Man, Ottawa, Canada, 1977). The latter book documents the simple, mostly outdoor, clay bake-oven tradition of Quebec and might not be ideal for the cellar application you want.

Another good source of information is the Oven Crafters' catalog (P. O. Box 24, Tomales, Calif. 94971; 707-878-2028). Alan Scott, the owner of Oven Crafters, has been doing oven workshops (mostly backyard models) for several years and has developed some plans.

As for heating the oven with your woodstove, you should know that New England traditional brick ovens were always heated by placing either coals or a fire directly in them. Hot smoke from a woodstove will create an underheated, dirty oven unless the stove is burned very, very hot.

The traditional bake oven consists of an interior brick dome with a small opening at the front. The dome is incorporated into and enclosed by the walls of the home's central fireplace. The floor of a traditional oven is made of stone, brick or firebrick laid dry on a bed of sand. A mortared floor will fail from expansion. The oven dome is usually built with one course of common brick or firebrick standing on edge "soldier fashion." The remaining bricks of the dome are laid flat and tipped slightly toward the center, with each course closing in tighter to form an igloo shape. This may necessitate cutting the bricks into a trapezoidal shape. To allow for expansion, clay mortar or fire clay should be used, not portland cement-based mortar. Ovens built with portland cement mortar will self-destruct in response to the heat.

An oven form can be built up with wet sand or with arched plywood sections covered with thin lath. Ovens can also be cast in orange-peel shaped pieces from castable refractory concrete on a sand form. Refractory concrete is available from refractory wholesalers all over the U. S. A 2300° to 2500° concrete would be appropriate. Because the dome will undergo expansion and contraction, its base should include an expansion joint with ¼ in. of gasketing. Mineral wool or high temperature fiberglass work well as gasket materials. A traditional method of insulating the dome against heat loss is to cover it with sand. But modern high temperature insulation allows the oven to be built in such a way that it need not lose heat as rapidly as a sand-covered or uncovered dome ovens want to do. Three or more inches of mineral wool or high temperature fiberglass can be used for this purpose as well.

To my knowledge nobody manufactures traditional American brick-oven hardware today. Old doors, however, are often saved by masons or antique stove dealers and can be found with effort. The so-called tall arched-top "bake-oven door" made in the southern U. S. and available at many masonry supply yards is not appropriate for a true low-arch traditional bake oven.

PAINTING STONE

We recently bought a house that meets most of our needs and is in a good location. However, we're not happy with the orange and tan hues of the stone that covers about three-quarters of the front of the house (the remainder is sided with cedar shakes).

We want to paint the stone, but a stonemason friend of mine warned me that the paint would never stay on because of the moisture in the stone. Is there a product we could prime the stone with that would help the finish coat adhere?

—Peter D. Robinson, Chattanooga, Tenn.

Stephen M. Kennedy, a stonemason in Biglerville, Pa., replies: If I had a stone wall (or any masonry wall, for that matter) that was ugly enough to deserve a paint job, I'd be inclined instead to parge the wall with a pigmented mortar mix. Parging is fast, cheap and fairly long-lasting if done right. (For more on parging and stuccoing see *FHB* #18, pp. 33-35. See also *The Plaster and Stucco Manual,* available from the Portland Cement Association, 5420 Old Orchard Rd., Skokie, Ill. 60077; 708-966-6200; $6.50 plus postage and handling.) And by using a pigmented mortar mix, the color will not peel off or wear out. The moisture is actually beneficial to the parge coat; in fact, parging should only be done on a pre-moistened surface.

Parging is like stuccoing, but with a little luck and some finesse it can be accomplished in a single coat. Most cement suppliers carry at least a few different colors of mortar mix and can usually order a much wider variety of colors, including custom colors. You can trowel the mortar onto the wall to give it a textured adobe look. Or, with extra finishing, you can leave it as smooth as a drywall interior.

If you need to match existing painted surfaces, I recommend parging the wall with a white mortar mix, then painting that surface after it has cured, usually within a few days. A mix of one part Type I white portland cement, two parts Type S hydrated lime and seven parts fine masonry sand should be about right for a good parge coat. Resist the temptation to use too much portland because that will cause the finished product to contract while setting, causing cracks in the finsh coat.

Victor DeMasi, a painter in W. Redding, Conn., also replies: In most cases stone can be successfully painted. However, two conditions can be particularly troublesome. The first occurs where the surface of the stone is highly polished. In this case it must first be abraded to provide tooth for the proper adhesion of the paint. The second occurs where excessive moisture migrating from the interior side of the wall causes premature failure of the paint. This happens most often in older houses lacking vapor barriers. If moisture from the interior isn't a problem, then a good dry spell is all that's necessary before painting. I would recommend using a high-quality latex masonry paint that allows the surface to breathe. Follow the manufacturer's recommendation on the primer. In many cases, thinning the first coat suffices.

You can also pickle the stone (see *FHB* #60, pp. 78-82). Thin the latex masonry paint before application to 1 part paint and 3 parts water. Brush it on as a very thin coat. Blot up excessive runs with a sponge. You can apply more than one coat to achieve the desired opacity. Pickling will soften the color of the stone, without obscuring its character. It also sidesteps the moisture problem. The thinned paint acts more like a stain than a paint, because it doesn't form a solid film that's subject to failure. In any case, minor areas of failure are of little consequence on such a variegated surface and might actually add to the charm. If you paint or pickle, I suggest you experiment with a small sample area in an inconspicuous spot before proceeding.

I've also successfully pickled brick. A light wash of white applied liberally can tone down the harsh red of some interior bricks. For exterior brick I've used white oil paint, thinned as above and applied without a primer. In time, the paint begins to peel off in very fine shreds, giving the surface a soft, slightly pink color as the brick underneath becomes exposed. The result is a beautiful aged appearance.

READERS REPLY

To Peter Robinson of Chattanooga, who asked for advice on painting the stone veneer on his house, you gave two responses. The first was, don't paint it, parge it. Yikes. Some advice—if you don't like the stone, just mud the house. Your other advice listed some painting tricks.

If it's not too late, let me join the advice team. Don't paint that stone! Clean it. That orange and tan stone is almost surely Tennessee Crab Orchard stone, which was used as a veneer on many houses built in the mid-South around the turn of the century. It's rare and costly now. Nashville is full of houses that are veneered with the stuff, and, like Robinson's house, they're trimmed with cedar shingles. Strangely, people seem lost as to what to do with these houses. Most people end up coloring their houses in a monochrome tan, roof and all, trying to make everything match.

My across-the-street neighbors have the mixed blessing of having historic restoration advisors for neighbors. They asked around and got some good advice: once Crab Orchard stone is clean, it's hard not to like. After a mild acid pressure wash, the stone revealed hues of chocolate brown, purple and pink, as well as the orange and tan that was showing through 70 years of dirt. My neighbors also got rid of their old tan color scheme. They painted the cedar shingles in the gables a medium gray-green (the shingles on these houses were almost always stained deep blue-green originally, but you can't duplicate the effect with paint), and the wood trim a light ivory. They also put on a grey roof. The house looks great.

So please, don't go telling people in Tennessee to paint or (gasp, choke) parge an unusual example of regional architecture. All they need is a contractor who can do a decent pressure wash and some paint colors that work with the stone.

—*Walter Jowers, Nashville Tenn.*

ISOLATION MEMBRANES FOR CERAMIC TILE

I want to put a ceramic tile floor over a T&G 2x6 subfloor. I would like to use a 1-in. thick mortar-setting bed. Michael Byrne states in his book *(Setting Ceramic Tile,* The Taunton Press, 1987) that it is necessary to use an isolation membrane between the mortar bed and subfloor. Typically, tar paper is used in this application. However, my wife is allergic to petroleum products so I need to use some other material. Is it possible to use Dennyfoil as an isolation membrane? This product is aluminum foil laminated to both sides of some type of paper.

—*H. M. Macksey, Olga, Wash.*

Bob Wilcoxson of Island Tile in Edgartown, Mass., replies: An isolation membrane (also called a cleavage membrane) isolates the tile from movements in the subfloor that could cause the tile to crack. Ideally, an isolation membrane should also provide some waterproofing. I called Denny Products (3500 Gateway Dr., Pompano Beach, Fla. 33069; 800-327-6616), which distributes Dennyfoil. They said that Dennyfoil is manufactured as a housewrap and is unsuitable for use as a tile membrane.

Although I often see tar paper used as a membrane, the Tile Council of America's handbook specifies 4-mil polyethylene for this purpose. One of my favorite membrane products, though (which is also one of Byrne's favorites), is a product called NobleSeal (The Noble Company, 614 Monroe St., Grand Haven, Mich. 49417; 616-842-7844). It is made from chlorinated polyethylene reinforced with polyester fiber. It comes in 60-in. by 100-ft. rolls and is marketed as a waterproofing and isolation membrane. NobleSeal costs a lot more than polyethylene sheeting, but its flexibility and water resistance make it worth the cost.

Another thing I'm concerned with is your T&G subfloor. Plywood is relatively stable; pine, on the other hand, has a much greater tendency to cup and bow. My concern is that a T&G subfloor—even one consisting of 2x6s—might move enough to crack the tile, membrane or no membrane. I wouldn't worry about this in dry climates, but it could pose a problem in a damp area like Washington state. If possible (if it doesn't wreak havoc with your floor heights), I recommend that you install an underlayment between your tile and your subfloor. My first choice would be a cement backer board, although you could use AC plywood secured with nails and a good construction adhesive.

REFINISHING PORCELAIN

Is it feasible to refinish porcelain enamel on cast iron? What about refinishing the glaze on vitreous china? I have some maroon Art Deco-era pieces that have large chips in them. The local firms that "refinish" porcelain spray a plastic finish over the enamel and glaze with results that look good at first but seem to be very short-lived. A friend masked off an iron wall-hung sink to paint his wall and pulled the new finish off the sink with the tape. Are there different products with different characteristics? Does anyone refire porcelain enamel on something as big as a tub?

—Joel Herzel, Santa Cruz, Calif.

Audrey Fontaine of Miracle Method Bathtub Refinishing, Ludlow, Mass., replies: All porcelain enamel fixtures have a glass-like surface that is fired onto the steel, iron or ceramic base in a kiln at thousands of degrees. The firing process results in a high-gloss, extremely hard finish that is ideal for using in bathrooms. Abrasive cleansers and wear will cause loss of this gloss, necessitating refinishing.

Proper preparation of the surface prior to refinishing is most important for the adhesion, and thus the life expectancy of the coating. The type of coating makes little or no difference in life expectancy. Most refinishers today use an acrylic urethane, which gives a very durable, long-lasting surface.

There are two basic methods of refinishing porcelain and vitreous china. One uses an acid etch to create a mechanical bond; the other uses a chemical bonding agent to create a molecular bond between the original porcelain or china and the new finish. Companies that use an acid etch generally guarantee their work for up to one year. We use the chemical bonding agent and offer a five-year warranty.

Our procedure is as follows. The fixture is cleaned thoroughly, using industrial-strength cleaners. Old caulking is removed. The drain and overflow pieces are removed. Damaged areas are filled to conform exactly with the original contours.

Next a bonding agent is applied. Because of the chemical bond, water will not work its way between the coating and the porcelain.

After application of the bonding agent, three coats of the highest-quality acrylic urethane enamel are sprayed on using low-pressure, hot-air equipment. Acrylic urethene is an extremely hard, chemically resistant coating that has been proven in industrial uses (on the exterior surface of jet aircraft, for instance). After spraying, an infrared heating system is set up to speed the curing. This takes six to eight hours. The last step is the application of a polymer glaze.

CONCRETE-BLOCK PRIVACY WALL

I want to have a "privacy wall" constructed between my neighbor and myself. I want this to be a concrete-block wall with stucco. It will be 100 ft. long and 8 ft. high. This is a non-bearing wall in sandy loam soil, with a water table about 2 ft. to 3 ft. below grade (typical for central Florida).

I have had different estimates ranging from $2,900 to $8,300. The higher quote was for a block wall with poured-concrete posts every 15 ft. and a poured-concrete lintel, which the contractor says is necessary. The other estimate was for concrete block with vertical rebar in the footer every 12 ft. Which is adequate for my needs?

—*Gregg Eads, Winter Park, Fla.*

Mason Dick Kreh of Frederick, Md., replies: The more expensive wall would be stronger, but, with a few changes, the less expensive one would suffice. You don't have much of a freeze problem to worry about in your area, so the footing and foundation for the wall would be mostly for load distribution. My recommendations for building a structurally sound concrete-block wall at the lowest cost are shown in the drawing on p. 206.

First, check with your local building inspector or department of permits to see what they require to conform to the code. I suspect that you need to pour a footing that is 8 in. deep and twice as wide as the block being used (in this case 8x8x16 block would be more than ample). Also, some rebar reinforcement should be placed in the footing as it is poured. About 2 ft. below grade should more than meet the code for your area.

A masonry wall 100 ft. long, like the one you describe, should be reinforced with vertical rebar in the block cells about 12 ft. o. c.

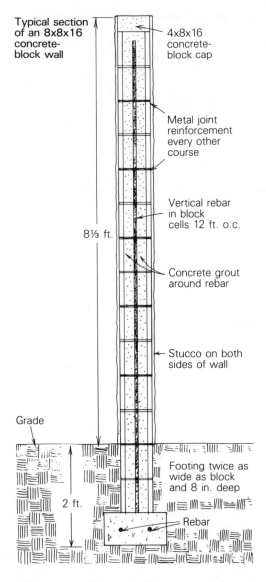

Typical section of an 8x8x16 concrete-block wall

4x8x16 concrete-block cap

Metal joint reinforcement every other course

Vertical rebar in block cells 12 ft. o.c.

Concrete grout around rebar

Stucco on both sides of wall

8⅓ ft.

Grade

Footing twice as wide as block and 8 in. deep

2 ft.

Rebar

Then fill around them with concrete grout, puddled down with a wood rod to compact it. I would also use metal-wire joint reinforcement every other course, or every 16 in. This is not expensive and will really strengthen the wall. Metal joint reinforcement is available from your local building supplier. Buy the 8-in. size to match the 8-in. block width you are using.

The type of mortar you use is especially important for a wall like this one. I recommend using Type S masonry cement mortar in the proportions of 8 shovels (dirt shovel) to a half-bag (16 shovels to a full bag) for best results. Your building supplier will also stock this. Type S has a much higher flexural strength and better bonding to the block than normal (Type N) mortar does. It has good compressive strength—at least 1,800 psi in 28 testing days—which is ideal for what you are doing.

I would cap off the wall with a solid concrete cap or 4-in. block that's the same width as your 8-in. block.

You mentioned applying stucco to the wall as a finish, which I know is common to your part of Florida. At most building suppliers, you can get bags of dry premixed stucco, and all you add are sand and water. You can also make your own stucco as follows:

Begin by mixing 1 part portland cement to 2 parts of mason's hydrated lime to 8 parts of sand with water.

The color of the sand being used will affect the color of the finished stucco. In your area, the sand has a lot of whitish materials in it such as coral and seashells, which will give you a white stucco.

Keep in mind that Florida sand is rather coarse and sharp so you may have to cut down on the amount to obtain a good workable stucco mix to plaster with. A little experimentation will soon give you the right recipe.

A quoted price depends on the contractor's overhead, his wage scale and the cost of materials in your area. I advise you to get about four bids before you make up your mind. You could then be assured that the price is fair. Or you could build the wall yourself and enjoy the fruits of your labor.

ATTACHING PLASTER CASTS

I would like to know what adhesives would be recommended for attaching plaster casts to drywall and wood.

—Pat Starck, New Hampton, N. H.

Plasterer David Flaharty of Green Lane, Pa., replies: On drywall, wood and sometimes even on plaster I use a mixture of white glue and plaster to attach plaster casts. White construction adhesives are also satisfactory, but my mix offers the advantage of allowing me to prepare small batches rather than puncturing a whole tube for a little job. Casts must be dry and the backs moistened slightly (I use a floral mister), which ensures a good bond with the white glue. This method works best on unpainted surfaces.

With lightweight casts (such as bellflowers) it is sufficient to apply the adhesive and press the ornament into place. With heavier casts I apply the adhesive, press the ornament into place and then remove it to check whether the adhesive has full coverage. I wait a minute before re-applying the cast, which allows both wet surfaces to dry a bit. The ornament must be held, or propped up, until it is secure. With heavy casts I pre-drill and countersink holes into which I drive coarse, threaded, galvanized screws. The screws hold the cast until the glue dries. Later, I point up the holes with plaster. Another method is to use galvanized finishing nails driven at angles through the casts.

PEELING PLASTER

Our company renovated a large hallway in a home built in the 1890s. After stripping off four layers of wallpaper, we patched about 15% to 20% of the total wall surface with ³⁄₈-in. drywall and joint compound. An additional 10% to 15% of the surface was mechanically reaffixed with screws and plaster washers, with the balance of the surface requiring only a skim coat.

All surface areas were coated with Durabond 90, followed by two coats of joint compound. No plaster bonding agent was used. Dry surfaces were wetted before patching, and a fast drying, alkyd-base primer was applied as a sealer.

The client chose an acrylic-coated, prepasted paper, hung to the manufacturer's specifications. One week after completion of the paperhanging, a 15-in. by 30-in. section of the original plaster separated from the ½-in. thick scratch coat. Later the client called to report that the seams of the wallpaper were lifting. Upon close inspection, it was apparent that both the Durabond and the original lime coat had lifted from the scratch coat along the wallpaper edges. The ceiling was not papered and shows no similar problems. No water damage was evident, and there did not seem to be any pattern to the delamination.

We have resecured one of the areas by injecting white glue into the seam and applying consistent pressure for 24 hours afterward. It seems to be holding but leaves a lot to be desired in appearance. We have been in the business of renovation and decorating for ten years and have never encountered a similar problem.

—*Michael G. Byrnes, Buffalo, N. Y.*

Dean Russell, an historic restorationist in Mattituck, N. Y., replies:
I think the problem resulted from the delamination of the original finish coat (white coat) of plaster from the base (brown coat). This is common in some older houses. The root cause may be any one of several possibilities. The brown coat may have been too "poor" a mix (too much sand) or the finish coat applied when the base coat

was too "green," or too wet. The brown coat may have set too slowly, allowing the surface to dry prematurely, creating a dusty, weak powder film on the surface. The white coat would only tentatively adhere to this "dead" plaster surface. The same would be the case if the brown coat plaster had been floated with too much water or troweled too smooth. Any of these errors in original workmanship would cause the finish plaster to peel.

Why did the plaster peel off now after all these years? Several factors might be considered. First, the scraping, chipping and other construction work caused some vibrations, which could have a tendency to loosen the plaster. Then, the application of the water-based Durabond 90 and the regular joint compound, which dampened the white coat, may have caused some expansion and contraction of the old finish plaster. Since the underlying brown coat remained relatively dry and stable, there would be unequal stress on each layer, further weakening the already inadequate bond.

Furthermore, since wallpaper glue is designed to shrink somewhat as the glue and paper dry, another contraction force went to work on the wall. The bond of the wallpaper glue to the properly primed wall surface became stronger than the finish plaster to base-coat bond. Because each section of wallpaper would shrink as a unit, the strongest pulling force would be focused along the edges of the paper, lifting plaster there. It seems the wallpaper may have been the last straw, because none of the peeling happened on the painted but unpapered ceilings.

The solution is not a pleasant one. The method of repair using white glue and pressure may be moderately successful, but I question its durability. The only sure way to rectify the situation is to remove the wallpaper and scrape off all the original finish coat. Wire brush the base coat to remove any soft or dusty surface material. Brush off the walls with a soft-fiber brush to remove the dust.

Next, apply a bonding agent/prewetting solution of 2 to 3 parts water, 1 part denatured alcohol, 2 parts bonding agent (Plasterweld, Thorobond or the like) and allow it to penetrate the plaster. This will, to some degree, consolidate the surface of the brown coat and will break the surface tension of any dry dust still on the surface. While still tacky, apply a thin, full-strength coat of bonding agent. When dry, plaster the walls with a "high-gauged" mix (more plaster to lime putty ratio) and trowel smooth. A variation might be to use veneer plaster, but a knowledgeable plasterer should be consulted to inspect the actual job-site conditions.

FLOOR FRAMING FOR FIREPLACES

I want to build two corner fireplaces, side by side in adjacent rooms, with two flues in one chimney. How do I frame the floor around the chimney opening?

—Stan Davis, Northville, Mich.

Scott McBride, contributing editor to Fine Homebuilding, *replies:* The floor framing around a catty-cornered fireplace requires a diagonal header of some sort. If the header is recessed, tail joists can simply bear on top of it with their ends strung together by a diagonal band joist. If a flush header is used, the joists will hang on the side of the header (shown below). In this case the diagonal cheek cut of the tail joists will afford a good nailing surface, but for

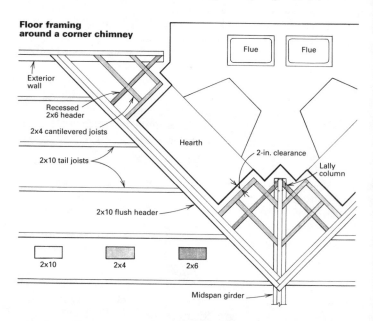

Floor framing around a corner chimney

Exterior wall

Recessed 2x6 header

2x4 cantilevered joists

2x10 tail joists

Flue

Flue

Hearth

2-in. clearance

Lally column

2x10 flush header

| 2x10 | 2x4 | 2x6 |

Midspan girder

extra strength you can use metal hangers skewed 45° to hang the joists (Simpson SUR/SUL series, Simpson Strong-Tie Co., P. O. Box 1568, San Leandro, Calif. 94577; 510-562-7775). Flush headers create equal shrinkage as the lumber dries. A recessed wood header can cause a dip in the floor as the header shrinks.

Codes vary, but a minimum 2-in. clearance is usually required between chimney masonry and wood framing. In most situations this means the header runs 2 in. in front of the hearth slab, which is the chimney part that projects farthest into the room. The hearth can be a trouble spot when framing around any chimney because it's usually narrower than the overall width of the chimney. This difference in width means that two corners of the floor opening must be extended inward without bearing on the masonry. The solution is to span the corner with a short, recessed header. Cantilevered joists can then bear on this recessed header and extend inward to reach within 2 in. of the hearth. The subflooring can then span this gap to butt against the masonry.

In the drawing on the facing page, one end of the double 2x header bears on the exterior wall, and the other end bears on a midspan girder between the two rooms. If the girder between the rooms is steel, it can bear directly on the chimney. (Some codes even allow wood beams to bear on a chimney, provided the beam is placed a minimum distance from the flue.) Alternatively, the end of a wood girder could be picked up by a short piece of steel channel spanning the inside corner formed at the centerline of the chimney. Another solution would simply be to post the girder from below with a Lally column.

SMELLING SMOKE

We have a tight, well-built house with two masonry fireplaces, one in the great room and one in the master bedroom. The problem is that the fireplace smell lingers for days after we have a fire, whether the damper is open or closed. At times, this smell is heavy enough in the bedroom to make sleeping and breathing uncomfortable. The smell may reach other parts of the house, but is most prominent in the rooms with the fireplace.

It's clear that even with the damper closed a strong downdraft occurs when there is no fire. In fact, when a fire is started, smoke comes into the room unless the flue opening is heated with burning paper first. When in use, both fireplaces function and draw beautifully, and generally the wood is completely burned when the fire is allowed to burn down overnight. Short of tearing out the fireplaces and starting over, is there any way to correct or at least minimize the problem?

—Leonard R. Bobbins, Cleveland, Tex.

Albert A. Barden, III, of Maine Wood Heat Co., Inc. in Norridgewock, Maine, replies: To answer your question, I consulted two master masons: Dale Hisler of Norwich, Vt., and Bob Gossett of Yakima, Wash. We all agree that some form of "negative pressure" causes the downdraft and smoke smell. Air from the chimney is falling (cold air is heavier than warm), or being sucked into the house, bringing with it the smell of the carbon deposits on the chimney flue liners. You mention that the house is tight, but you don't mention any make-up outside air for either fireplace, or for other heating systems in the house.

It may be that when other appliances are in use, such as an oil or gas furnace or a bathroom exhaust fan, negative pressure is created in the house that draws supply air through the two fireplace chimneys, bringing with it the smoky smell.

Hisler typically installs make-up air for fireplaces in the baseboard near the fireplace. He also retrofits standard fireplace dampers by grinding them down smooth and adding a fiberglass rope gasket with silicone caulking to make the dampers airtight when closed. Gossett pointed out that glass doors added to fireplaces often alleviate the post-fire smells. Tighter dampers and new air supply for other household appliances should solve the problem.

DAMPER FOR A RUMFORD FIREPLACE

Neither *The Forgotten Art of Building a Good Fireplace* by Vrest Orton (Yankee, Inc., Main St., Dublin, N. H. 03444; 1974) nor *Architectural Graphic Standards* (American Institute of Architects, John Wiley & Sons, Inc., 605 3rd Ave., New York, N. Y. 10158; 1988) provides an adequate explanation of the damper for a Rumford fireplace. Orton does caution that the throat should be narrower than is customary with modern fireplaces, and that if a conventional damper is used, "the mason should be instructed to bury the front frame of the damper in the masonry." Other sources caution that installation of the damper must allow for expansion when heated.

How do you keep the masonry below the front frame (which narrows the throat) from cracking and coming loose? A letter from R. W. Heslop on p. 8 of *FHB* #47 says he's built two. How did he solve the damper problem?

—*Dirk Van Duym, Mandeville, La.*

R. W. Heslop replies: The question of damper installation gave me pause also. I finally concluded that the slope at the rear of the breast didn't make the fireplace heat better. This meant that it could be vertical, not slanting, up to the damper (see the drawing on the facing page). At the damper, the wall was increased in width to two bricks. I used a cast-iron damper purchased at the local building-supply outlet that matches, in length, the opening chosen for the fireplace. The front frame of the damper is set against the rear face of the breast. Mortar and brick were laid onto the top front slope of the damper as the width of the breast was increased. The ends and rear frame sit on, not in, a mortar bed.

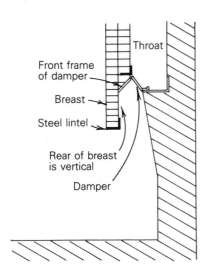

Front frame of damper

Throat

Breast

Steel lintel

Rear of breast is vertical

Damper

They are on top of the coving and fire back and are free to move for expansion and contraction. By installing the damper in this manner, the front half of it, in effect, becomes a lintel when the wall is widened. Since I am not professionally trained, I was not sure this would provide sufficient strength, so a second lintel was installed above the damper.

The damper has several notch settings to control the size of the opening or throat. These openings range from about 1 in. to more than 4 in. The first fireplace has been in service 19 years, the second one, six years. Both perform flawlessly.

SIZING A FLUE

Since moving to northwestern Montana, I've had trouble getting chimney flue linings larger than 16 in. by 16 in. I've heard of putting two linings together side by side, creating two flues for one fireplace. Would a pair of flue linings work? Also, I have a copy of a Los Angeles chimney code that shows solid grout around the flue lining, yet I was taught to leave an expansion area of ½ in. to 1 in. around a flue lining. Which is the correct method?

—Jim Cutting, Somers, Mont.

Carl Hagstrom, a builder and mason in Montross, Pa., replies: The area of a flue lining should be at least ¹⁄₁₀ the area of the fireplace opening. If a 16-in. by 16-in. flue lining is the largest available in your area, that doesn't necessarily limit your fireplace opening to

17½ sq. ft. (10 times the 16-in. square flue lining). If your fireplace is designed with a larger opening, you can use a double flue.

The National Fire Protection Association (NFPA) allows two flues to vent a single fireplace, and sizing them to meet the ¹⁄₁₀ rule should be adequate. But from my experience, it never hurts to oversize the flue area in a fireplace.

The NFPA also requires the flue lining to be separated from the chimney wall by an unobstructed airspace that's at least ½ in. wide but no more than 1 in. wide. You'll have to separate the pair of flue linings, too. To seal off the area between the two linings, rest them on a piece of ½-in. by 3-in. flat steel stock built into the smoke chamber. Flue linings experience large thermal swings, and the space around them allows for expansion without affecting the chimney walls. Never fill this space with grout. When laying up the courses of flue lining, use a nonwater-soluble refractory cement, but use only enough to make a good joint; the refractory cement should squeeze out just enough to hold the flue linings in position.

TESTING AND CONTROLLING ASBESTOS

The hot-water pipes in my basement are covered with a white insulating material that I assume is asbestos. Although most of it is intact, a few small areas are slightly damaged. What is the criteria for asbestos removal? Are there any testing kits that evaluate the level of contamination, like those for radon? Are there any commercially available sprays or coatings that provide a shield around the damaged areas?

—S. F. Lenahan, Pt. Pleasant Beach, N. J.

Stephen Flynn, a renovator and industrial hygienist from Lempster, N. H., replies: When asbestos-containing material (ACM) is disturbed, dangerous levels of invisible dust can be released. Exposure to this dust can cause cancer and other diseases. The risk isn't necessarily dose-dependent, either—a small exposure can still cause disease.

All aspects of ACM management (maintenance, removal, transportation and disposal) are regulated by federal, state and local laws. Most experts recommend hiring a certified professional to do this work. Your state department of environmental affairs or your local health department can tell you how to find one. Some states, however, let you legally perform certain ACM projects yourself. The agencies mentioned above can tell you more about the laws for your area. New Jersey, for example, lets you disturb up to 25 ft. of asbestos pipe covering.

If you want to do your own testing, you'll need to learn more about ACM management. Books and videos on the subject are available from UBIX Information Services (P. O. Box 1018, Haddonfield, N. J. 08033-0594; 609-429-4483). You might also borrow a book or video from the physical plant manager at your local school. This material should define the proper respirator as well as how to test fit it. I cannot overemphasize the necessity of test fitting a respirator. The fine asbestos dust can easily get past the edges of an ill-fitted respirator.

Testing for asbestos—To test your pipe insulation, choose at least one pipe elbow and one place a few feet away from an elbow. The covering used on straight pipes may be different from that on the elbows.

At this point, don the fitted respirator and some thin latex gloves. Make sure that nobody else has access to the basement. Fill a spray bottle with a solution of one teaspoon of dish detergent to a quart of water and gently mist the test site. This can take some time, as some asbestos-containing material is hydrophobic and does not wet readily. After dampening the site, cover a disposable cake pan with a small plastic trash bag and hold it under the site while using a new razor blade to cut a dime-size hole in the material. Re-mist the area if you expose any dry material. Then gently pry the cut piece out onto the pan. If the material appears dry, mist it; if even a slight puff of dust appears, mist the air around the cut to direct the dust into the pan. After you've cut the opening, use tweezers to probe the hole all the way down to the pipe. Your goal is to get a sample from each layer of insulation (some insulation material has asbestos in certain layers only). Place the sample in a plastic locking-top bag and label the bag (for instance, "elbow 1"). Then fill the hole with spackling.

Turn the small plastic bag that was covering the pan inside-out and place the pan inside. Then place the wrapped pan in a heavy plastic trash bag. Spray your gloved fingers so that the water washes any fibers into the trash bag, and add enough paper towels to the large trash bag to absorb any freestanding water. When you're completely finished, pull the misted gloves off and drop them into the large trash bag. Properly label this bag and store it in a safe place. It should be disposed of when any asbestos is removed.

Send the labeled bags to a laboratory. The one I prefer is CBC Environmental Services (140 East Ryan Rd., Oak Creek, Wis. 53154-4599; 800-365-3840). They'll charge you $10.50 to check each sample. The lab prefers dry samples; wet samples make the lab work a bit harder but also make the sampling procedure much safer for you. If you use a local lab, make sure that it's EPA certified. Your state agencies may have a list of approved labs.

Repair—Let's assume that both of your samples were found to contain asbestos. You're right to assume that undamaged material should be left alone. Small areas of damage can be easily patched. If the damage consists of small holes less than ⅛ in. deep, you could simply fill them with spackling. If they're deeper, you'll need to patch them. You can patch the holes by misting them and wrapping them with duct tape. The tape should extend past both sides of the wetted hole to dry, undamaged material. A better method is to cover the area with a product called Rewrap (International Air Filter, 200 N. Spring St., Elgin, Ill. 60120; 708-742-3900). This is a plaster-impregnated canvas that sells for about $1.50 per sq. ft. Regardless of the repair method, you should coat the repaired area with latex paint. The repaired pipe will then be at least as resistant to damage as the original.

But patching is only half the story. Chances are that dust has escaped from the damaged area and contaminated the nearby area. Unfortunately, there are no reliable testing kits available to nonprofessionals. I prefer the common-sense approach. If the damage is minor, the contamination may be minor. If you notice a few small lumps of material under the area you've patched, you can mist them and transfer them to your trash bag. You may also want to wipe the area under and around the pipe (including nearby walls or shelves) with wet paper towels and wet-mop the floor. A coat of paint will lock down any remaining fibers.

If you do attempt decontamination of asbestos dust, never use a vacuum cleaner. The invisible micron-sized fibers will pass through a common vacuum cleaner easily.

Removal—If the material tests positive and there's a great deal of damage, then removal may be the only option. Asbestos removal and the subsequent decontamination can range from easy to nearly impossible. It should only be performed by certified professionals.

In my experience, however, reliable and conscientious contractors can be hard to find. Use your state sources and check the references of recommended contractors. I strongly recommend hiring an independent industrial hygienist to oversee the removal project. Your hygienist should be EPA certified as a "project monitor," should have completed a NIOSH 582 equivalency course for testing and sampling asbestos, and should be equipped with a microscope for determining asbestos levels.

ROLLING EXTERIOR SHUTTERS

I am designing a new home to be built next year and would like to incorporate a feature I've seen on many houses in Germany. I believe the Germans call them "Rolladens." They are a metal window shutter that is built into the exterior wall of the house above the window

opening. The shutters are a tambour-type design and appear to roll up and down inside the wall. You operate them from inside the home by a strap that's also built into the wall.

These shutters seem to offer outstanding security without restricting egress during a fire the way burglar bars do. In a bedroom, they could be lowered to block out any offending light or street noise that normally gets through other window shades. The shutters would be a natural for homes built along the Gulf Coast, where windows frequently have to be boarded up for storms. Why haven't I seen these shutters in the United States? Do you know of a supplier?

—Irvin Bennett, Culpeper, Va.

Kevin Ireton, editor of Fine Homebuilding, *replies:* I can't tell you why rolling exterior shutters aren't as common here as they are in Europe, but I can tell you that they are available from Roll-a-way Insulating Security Shutters (10597 Oak St. N.E., St. Petersburg, Fla. 33716; 800-942-3230). Roll-a-way shutters are available with either PVC or aluminum slats, in a variety of colors and finishes. They can be operated either manually or electrically. And according to the manufacturer, the shutters increase the thermal efficiency of windows by creating an insulating dead-air space.

TIMBER-FRAME SOUND INSULATION

I am building a timber-frame house and wish to reduce sound transmission between the first and second floors. I was planning to adapt the floor system that Tedd Benson describes in his book, *The Timber-Frame Home* **(The**

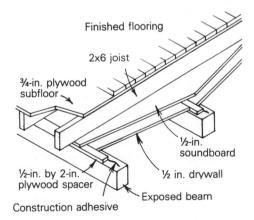

Finished flooring

2x6 joist

¾-in. plywood subfloor

½-in. soundboard

½-in. by 2-in. plywood spacer

½ in. drywall

Exposed beam

Construction adhesive

Taunton Press, 1988). From bottom to top, my deck (see drawing, p. 217) would include beams, drywall between plywood spacers, soundproofing board, 2x6 joists, ¾-in. plywood and finished flooring. I hope that this will give me adequate space for mechanical systems along with lessened sound transmission. For sound-deadening I was planning to use ½-in. soundboard.

After reading Russell Dupree's article "Quiet Please," *(FHB* #58, pp. 54-57) though, I am undecided about the effectiveness of this method. Is soundboard worth the extra cost and labor if used as described? Other than sand loading, which doesn't appeal to me, what alternatives are there?

—Elmer W. Kamm, Fort Wayne, Ind.

Author Russell Dupree replies: The floor details you describe will yield only minor improvements in both impact and airborne noise reduction over placing the T&G flooring directly on the exposed beams. Luckily, your design incorporates the wide (6-in.) air cavity, which you plan to use for mechanical runs. If used correctly, this cavity can also provide a tremendous noise-reduction benefit.

I suggest you construct a floating floor on top of the exposed-beam and drywall ceiling (drawing, facing page). This system uses small neoprene rubber "waffle" pads under the 2x6 joists to minimize noise transmission between the finished floor and the ceiling. Mason Industries makes a 2-in. by 2-in. by ¾-in. "Super W. Pad" specifically for this purpose (Mason Industries, 350 Rabro Dr., Hauppauge, N. Y. 11788; 516-348-0282). Each pad will support 8 to 10 sq. ft. of floor area. The rubber pads structurally decouple the floor and the ceiling, allowing the 6-in. air cavity to act as an acoustical spring, effectively dampening sound transmission (especially at low frequencies). When you have good structural decoupling, the addition of sound-absorptive materials, such as fiberglass, provides further noise reduction. This floor assembly will have sound-transmission class and impact insulation class ratings both around 45, which means that loud speech will be barely audible between rooms.

The term "soundboard" usually refers to lightweight wood or mineral fiberboard, and is often misleading. There are, unfortunately, no magic materials in noise control. Concocting a sandwich of "soundboard" and other materials often leads to disappointing results, frustration and a lot of expense. These fiberboard materials do have their use in reducing impact noise in certain situations, but they are not very effective in reducing airborne noise.

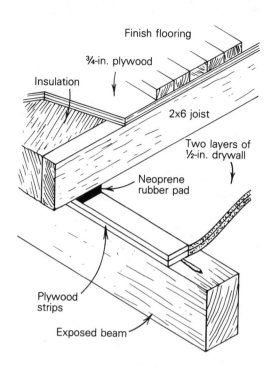

Finish flooring

¾-in. plywood

Insulation

2x6 joist

Two layers of
½-in. drywall

Neoprene
rubber pad

Plywood
strips

Exposed beam

You should also bear in mind that building a noise-resistant floor/ceiling construction will be to no avail unless you control the other sound-transmission paths as well. Don't punch holes in your floor or ceiling, and don't let sound escape through the door and down the stairs.

INDEX